Biotechnological Processing Steps for ENZYME MANUFACTURING

Editor
R.C. Tripathi

GENE TECH

2006
Gene-Tech Books
New Delhi - 110 002

ISBN 81-89729-01-2

Published by : GENE-TECH BOOKS
4762-63/23, Ansari Road, Darya Ganj,
NEW DELHI - 110 002
Phone: 41562849
e-mail: genetechbooks@yahoo.co.in

Printed at : Tarun Offset Printers
Delhi

PRINTED IN INDIA

Biotechnological Processing Steps for ENZYME MANUFACTURING

Preface

The application of modern biotechnology in enzyme manufacturing is dramatically sparking the exploitation of new enzymes and the development of new enzyme properties. Due to new technologies, new enzymes not accessible before can be cloned into and produced from a well-known host organism. Thereby, enzymes from almost any source in nature become accessible, including enzymes exhibiting unusual properties, such as extreme thermostability.

Applying new technologies, the enzyme properties may be efficiently altered which will lead to an increase in the variability of enzymes available and might lead to enzymes not present in nature so far. This mainly concerns enzyme stability, catalytic mechanism, substrate specificity and range, surface activity, folding mechanisms, cofactor dependency, pH- and temperature optima, kinetic parameters. Furthermore, enzymes could be chemically modified.

With applying new technologies, the variability in enzyme structure is dramatically increased and enzyme properties are significantly enhanced. Thus, these methods are mainly contributing to technical and economic goals. However, the safety of enzyme manufacturing might also be improved by restricting to few well-known and safe-to-use production strains which are used as hosts for genes from various sources.

The applications of biotechnology in enzyme manufacturing is elaborately described in this volume. It is hoped that this authentic and up-to-date work will address the needs of practitioners and students in the fields of organic chemistry, biology and bioengineering.

Editor

Contents

1

Introduction

The biological systems may be microbial (viruses, bacteria, fungi, yeasts, protozoans, algae), animal or plant cells or components of these cells such as enzymes or gene sequences. Up to now, the major workhorses of biotechnology have been the microbial cells, in particular bacteria and yeasts. However, one should not ignore the advantages and potential of using other cell types or systems.

Enzymes in Biotechnology

Enzymes are biological catalysts. With the exception of ribozymes, enzymes are mainly proteins in nature, and may also possess lipid or sugar moieties. Their main task is to decrease the activation energy (Eact) of a chemical reaction. The whole biotechnology field depends on the activities of enzymes, either acting alone or in concert.

Exploitation of enzymes is not a recent development. They have been used throughout the ages in leather tanning, in cheese-making, in the preparation of malted barley for beer brewing & the leavening of bread. These processes use enzymes in the form of whole cells. Conceptually, it is easy to decide whether one should use whole cells or isclated enzymes in a biotechnological process to produce or transform compounds.

Some processes require multi-step transformations, eg. the synthesis of antibiotics, interferons, monoclonal antibodies, or ethanol production from glucose. These transformations involve a number of enzymes acting sequentially, along with the requirement for co-factor regeneration in several steps. They are clearly candidates for using whole cells.

There may be some drawbacks to using whole cells in catalysis. These include:

— Competing side reactions from other enzymes.

— Sterility problems often arise when one deals with whole cells.

— Some conditions may lead to cell lysis, and this can result in loss of biocatalysts or severely decrease reaction rate depending on the nature and type of enzyme reaction.

— Formation of by-products.

On the other hand, with one-step or two-step stereospecific transformations, enzymes can be superior because their use will eliminate some of the potential problems with using whole cells. The advantages of using enzymes include:

— High catalytic rate possible.

— Functions in moderate conditions.

— Non-polluting catalysis.

— Less by-products formed.

— Less problem with sterility.

— Optimizing 1 or 2 enzyme catalyzed reactions is easier than optimizing the whole pathway in a whole cell.

— Enzymes catalyze a wide range of reactions, some with high stereospecificity, such as oxidation, reduction, dehydrogenation, dehalogenation, hydration, dehydration, etc. Therefore, the scope of enzyme technology is very broad.

Despite all these striking characteristics, enzyme have not been used widely as compared to chemical catalysis. This is because the use of enzymes also suffers from some serious drawbacks:

— Most isolated enzymes are not stable enough for use under industrial conditions. This may include poor stability with respect to temperature, pH, the presence of metal ions, etc. This is not surprising considering that most enzymes have evolved to function in a biological milieu where the conditions may differ from those used in industry.

— In cells, all enzymes are subject to turnover, ie. old ones are replaced by new ones. This is difficult to do with isolated enzymes in vitro.

— Most enzymes are water-soluble, and therefore, may be difficult to separate from reactants or products.

— Many useful enzymes are intracellular, and hence are difficult to isolate.

Immobilization

One approach to address the first two problems is the development of immobilized enzymes in the 1960s. The word "immobilized enzyme" was coined by Katchalski-Katzir in 1971. Enzymes may be immobilized on solid carriers by various techniques, such as carrier-binding, cross-linking or entrapment.

A variety of carriers have been tested with different enzymes. Some of these include alginate, magnetite, *kappa-*

carrageenan, polyurethane, etc. There is sufficient variety to accommodate almost all available industrial enzymes. Interestingly, there are very few detailed studies to compare the efficacy of various immobilization methods or immobilization supports.

It is generally believed that the best support for one enzyme may not work as well for another enzyme, and that each enzyme ought to be evaluated to determine the optimal carrier matrix system and conditions. Generally, the decision will depend on (a) the various characteristics of the enzymes; (b) requirements of the specific application, ie. the operational conditions; and (c) the properties and limitations of the support system.

Enzyme immobilization is aimed to restrict the freedom of movement of an enzyme. When one wants to do enzyme immobilization, one must select the carrier (support, matrix) and decide on the method of immobilization. Carriers or support may be selected on the basis of the following considerations:

— *Physical properties*: strength, available surface area, shape or form (eg. Beads, sheets, fibers), compressibility of carriers, porosity, pore volume, permeability, density, flow rate, pressure drop.

— *Chemical properties*: hydrophilicity, inertness to enzyme(s), substrate(s) or cofactor(s), available functional groups for modification, ability to be regenerated or reused. Compatability with certain buffers ? eg. alginate is not compatible with phosphate buffer.

— *Stability*: stability on storage, residual enzyme activity on storage, mechanical stability of support material when subject to pressure or water flow.

— *Resistance*: against bacterial or fungal attack, disruption by chemicals, pH, temperature, organic solvents, enzymes such as proteases.

— *Safety*: toxicity of component reagents, health and safety for process workers and end product users. Carriers must be "safe" if the end product is to be used for food, medical or pharmaceutical applications. For example, acrylamide is toxic.

— *Economic*: Availability and cost of carrier materials, chemicals, special equipment, reagents, technical skills required, industrial scale chemical preparation, feasibility for scale-up, continuous processing, effective working life, re-usability.

— *Reaction*: any diffusion limitations on mass transfer of cofactors, substrates or products ? Side reactions ?

— *Biodegradability re: waste treatment*: eg. polyurethane is not naturally occurring and is not easily degraded. Therefore, its use may be undesirable for cleanup.

There are several methods of immobilization, as listed below. Methods compatible with the enzyme, substrate or cofactor should be selected. For example, the entrapment method may not work well with cellulose substrate because the substrate itself is large and will not readily go into the entrapment matrix to reach the enzyme.

Adsorption

In adsorption, the enzyme is bound to the carrier material via reversible surface interactions. The forces involved are electrostatic, such as van der Waals forces, ionic and H-bonding interaction, and possibly hydrophobic forces. The forces are generally weak, but there are sufficiently large to allow reasonable binding. The carriers with adsorption properties are selected on the basis of knowledge of their compatibility with the enzyme. Adsorption utilizes existing surface interactions between enzyme and carrier, and does not require chemical activation or modification. Normally, one does not see any damage to the enzyme by the carriers.

Some advantages of adsorption include:

— Little or no damage to the enzyme.

— Simple, cheap and immobilization can be done quickly.

— No chemical changes to carriers or enzyme.

— Easily reversible.

Some disadvantages of adsorption include:

— Desorption or leakage of enzyme from support. Desorption may occur on changing environmental conditions (such as pH, temperature, ionic strength) or on conformational changes arising from substrate/cofactor binding, contaminant binding, etc. Some physical factors such as flow rate, agitation by stirring, collision or abrasion can cause desorption.

— Non-specific binding by substrate, cofactor or contaminants to the carrier may result in diffusion limitations and mass transfer problems. This may in turn change the kinetic properties of the reactions. Binding of protons to the support may change the pH of the microenvironment and change the reaction rate.

— Overloading of carrier possible. This may result in reduced catalytic activity.

— Possible steric hindrance by the carrier material.

Covalent Binding

This involves the formation of a covalent bond between the enzyme and the carrier material. The bond is normally formed between functional groups on the carrier and the enzyme. Those on the enzymes are usually amino acid residues such as amino (NH_2) group from Lys or Arg, carboxyl (COOH) group from Asp, Glu, hydroxyl (OH) group from Ser, Thr, and sulfhydryl (SH) group from Cys. Tests must be run to ensure the formation of covalent bonds will not inactivate the enzyme.

Through chemical modifications, functional groups on the carriers can be altered to different forms to accommodate different kinds of covalent bonds to be formed with the enzyme. For example, chemical modification of -OH group can give rise to AE-cellulose (aminoethyl), CM- cellulose (carboxymethyl), and DEAE-cellulose (diethylaminoethyl). This increases the range of immobilization methods that can be used for a given carrier.

While many supports exist, an important factor for enzyme immobilization appears to be hydrophilicity, which helps to maintain enzyme activity in a hydrophilic milieu.

Entrapment and Encapsulation

In both of these methods, the enzyme remains free in solution, but restricted in movement by the lattice structure of a gel. The porosity of the gel lattice is controlled to ensure that the structure is tight enough to prevent leakage of enzyme, while allowing free movement of substrates, cofactors and products. Encapsulation generally refers to larger capsules that allow for co-immobilization of different combinations of enzymes for selected applications. Notably, red blood cells can be used as encapsulation capsules.

Crosslinking

This method joins the enzyme to each other to form a

large, 3-D structure without any support. This can be done through chemical (covalent bonding) or physical means (flocculation). Tests must be done to ensure the active site remains free and available for catalytic activity.

Encapsulation often improves the operational stability and sometimes efficiency of enzyme catalysis under industrial conditions, and also allows their easy separation from the medium. This allows for easier recovery of the enzymes for repeat use. The improved stability and ease of separation of immobilized enzymes make them suitable to be used in a variety of bioreactors. They are also easier to scale up. In addition, encapsulation allows for ease of handling and storage of enzymes. Sometimes other properties of the enzyme may be altered by immobilization, eg. altered pH-activity profile.

There are also some disadvantages to enzyme immobilization. In some instances, enzyme activity may be reduced by immobilization. There can also be diffusional or mass transfer limitations arising from immobilization. Immobilization also means additional costs. Despite these disadvantages, it is generally believed that the use of immobilized enzymes will increase in the future in various industries.

Catalysis in Organic Solvents

Another approach that has generated some interest in enzyme technology is catalysis in non-aqueous or nearly anhydrous media. Traditionally, enzyme reactions are carried out in aqueous media. The question arises as to how much water is really needed by the enzyme to function. This led them to test how enzymes function in various amount of organic solvents. They found startling changes to enzyme properties as the water content of the reaction mixture decreases. Some of these are illustrated in overheads for porcine pancreatic lipase. They concluded that it is the water bound to the enzyme that is critical for activity, rather than the total water content of the system. Research by Klibanov, his colleagues and other researchers have opened up novel opportunities in enzyme technology and this field has changed from a mere curiosity to a serious alternative to chemical reactions.

The functions of a number of enzymes under non-aqueous conditions have been studied. Such systems are applicable to enzymes of varying reaction types. Some examples include:

— Lipases, which show increased thermostability, altered substrate specificity and a propensity towards synthetic reactions in non-aqueous media.

— Horse radish peroxidase (HRP) which is able to depolymerize lignin in 95% dioxane.

— Glucose isomerase for conversion of glucose to fructose to produce high fructose corn syrup (HFCS). - reference not available.

— Other examples include chymotrypsin, alcohol dehydrogenase, etc., ie. a variety of enzymes from different classes have been shown to function in non-aqueous media.

Some advantages to using enzymes in non-aqueous solvents include:

— Increased solubility of nonpolar substrates.

— Shifting thermodynamic equilibria to favor synthesis over hydrolysis.

— Reduction in water-dependent side reactions.

— Immobilization is unnecessary, because enzymes are insoluble in organic solvents, and can be recovered by simple filtration.

— Ease of product recovery from low boiling organic solvents.

— Enhanced thermostability of enzymes; water is required to inactivate enzymes at high temperature. Therefore, in the absence of water, enzymes are more stable.

— Altered activity and affinity (Km) of the enzyme.

— Novel enzyme reactions possible (eg. new catalytic activity seen with lipases or horse radish peroxidase).

— Elimination of microbial contamination.

— Can use enzymes directly within a chemical process.

Protein Engineering

One approach to address the stability problem of enzymes is to systematically alter the properties of enzymes by protein engineering methods. This requires the availability of a cloned gene and it would be helpful to have available the 3-D structure of the enzyme, resolved by X-ray crystallography. In the absence of detailed 3-D structure, one can do random mutagenesis and/or directed evolution. Most of these approaches will be covered later in the course. Most of the studies on protein engineering have dealt with improving the stability of an enzyme (stability with respect to chemical oxidation, temperature, pH). Other aspects include altering the substrate specificity, the cofactor requirement (NADPH to NADH) or imparting novel properties to a protein (eg. adding a Mn binding site into horse heart myoglobin at UBC).

Genetic Engineering

Genetic engineering can be used to improve production and purification of recombinant enzymes, eg. targeting enzymes for extracellular secretion to allow for their easy isolation and purification.

Examples of industrially useful enzymes include:

A. *proteases*: subtilisin (detergent enzyme), rennet (cheese-processing).

B. *carbohydrases*: amylases (starch hydrolysis), cellulases (bioconversions by Stake Technology), xylanases (bioconversions and biopulping), glucose isomerase (HFCS)

C. *pharmaceutically important enzymes*: urokinase (plasminogen activator) is a protease for the treatment of thrombosis and embolism.

D. *ligninases and Mn peroxidases*: pulp processing.

E. *restriction endonucleases*: used extensively in research.

Animal cells in Biotechnology

Animal cells are used in biotechnology mainly because of their level of complexity which allows them to carry out specific functions and express unique products which simpler organisms such as bacteria or yeasts are not able to do well yet. Almost all of the compounds of commercial interest made by animal cells are classified as high value, low volume products.

Animal cells can be grown as discrete cells, pieces of tissues, or whole or parts of organs. Explants from animals are generally referred to as organ or tissue cultures. If the connective tissues are disrupted by chemical, mechanical or enzymic (trypsin, hyaluronidase or collagenase) means, then a cell culture is obtained. Primary culture refers to the cells of tissues taken directly from the animals. After the first passage (subcultivation), a normal cell line is established. A normal cell line can become permanent (immortalized or continuous) if cells possess the potential for unlimited number of passages *in vitro*. Generally, after 70 passages at intervals of about 3 days each, a cell line is regarded as continuous. A transformed cell line refers to one where the cells become immortalized through the specific acquisition of foreign genetic materials.

Primary animal cell lines generally

— are diploid.

— are cultured as monolayers, anchorage dependent.

— exhibit contact inhibition.

— do not survive repeated subcultures.

Continuous cell lines generally

— are aneuploid (heteroploidy).

— are anchorage independent, grown in submerged cultures.

— do not exhibit contact inhibition.

— multiply more rapidly and to higher cell densities than primary cell lines.

— can be cultured indefinitely.

— may be smaller in size than primary cells.

— may exhibit decreased serum dependence.

— are better characterized compared to primary cell lines which may differ from preparation to preparation.

Biotechnological processes using normal or continuous animal cell lines have several disadvantages as compared to microbial cultures:

— Animal cell cultures are more costly to operate, due mainly to the requirement for complex and expensive media.

— Animal cell cultures generally yield lower cell density, commonly around 5x106/mL. This results in lower volumetric productivity (g product formed/L/h), specific productivity (g product formed/g cell/h), lower concentration of products accumulated & lower product yield (Y, g product/g substrate). This problem is being addressed by improvements in fermentation conditions to allow the use of dense cultures in the order of 108 to 109 cells/mL.

— Animal cell cultures typically exhibit slower growth rate and metabolic rate compared to microbial cells. The doubling time of animal cells is in the order of 20 to 100 h, with averages ranging from 18 to 24 h, compared to 0.5 to 2 h for most bacteria.

— Animal cell cultures are more susceptible to contamination by microbial cells, especially mycoplasmas which are difficult to detect and highly contagious.

— Many animal cell cultures require more careful handling to avoid cell breakage. This problem is being addressed by selecting for and using more robust cell lines. Most of the animal cells have to be grown on flatglass or plastic surfaces as monolayers; or as suspended cells in slowly rotating containers in gently agitated media.

Media and culturing (fermentation) systems are the two major factors that can seriously affect cost & productivity and render animal cells less attractive to use in comparison with microbial cells. Therefore, it's not surprising that they are the major foci of R&D with respect to production of biologicals from animal cells.

At one time, it was thought that genetic engineering can resolve many of these problems by transferring the necessary genes encoding the production of useful compounds from animal cells to microbial hosts. If so, genetically engineered microorganisms might replace animal cell cultures as production systems. However, this scenario has not happened. There are still many problems to be worked out with the use of recombinant microbial strains to produce heterologous proteins. Some of these include:

— Improper processing. Some of the animal cell products are not processed properly after synthesis by recombinant microorganisms. These include posttranslational modifications such as glycosylation. Glycosylation is assumed to be important for the activity of some animal cell products such as interferons.

— *Improper folding.* The correct folding of the heterologous proteins might not occur if produced in microbial hosts. With many pharmacologically important animal cell proteins, proper 3-D folding is essential for activity/function.

— *Insufficient types produced.* For some products, such as interferons, several genes or genetic elements (eg. promoters) are required for the production in animal cells. With recombinant organisms, not all these genes are introduced to the microbial hosts. For example, production of -IFN in human cells yields a final mixture of 12 different types. Each type has different chemical and biological properties. Interaction among the 12 types is assumed to be important in the overall activity of interferon. However, the ones produced by recombinant microorganisms only represent 1 or 2 of these types.

— *Intracellular accumulation.* With recombinant bacterial hosts, extracellular production of heterologous proteins can be a problem, although this problem is being studied. Intracellular accumulation of foreign proteins can occur in inclusion bodies or periplasmic space from which it is difficult to recover active functional proteins.

— *Stricter health regulations.* Health and safety regulations can be more strict with recombinant products. This results from the need to remove any microbial contaminants such as endotoxins, which can cause contamination and affect purification of the desired compounds.

Medium Formulation

One of the major considerations in using animal cell lines is the medium formulation. The medium is important because of its high cost and its effect on growth and productivity of the cell line. Most animal cells are not independent. They require growth factors from other cells in order to survive. These factors are usually provided in serum. The basic cell culture medium consists of 30 to 50 ingredients. Various supplements are usually added to the basic medium. These include glucose, serum (added to 5 to 10%) and antibiotics to give a complete growth medium. The basic medium, if purchased in bulk, costs about the same as the general bacterial media. However, the complete growth medium can cost 3 to 10 times more than bacterial media. On scale up, this cost factor increases dramatically, and the cost of the medium can become a high percentage of the total production cost.

Among components of the complete medium, serum and antibiotics are responsible for the bulk of the cost. For example, serum can account for 17 to 84% of the total cost of the growth medium components, while antibiotics can account for anywhere between 9 to 44% of the total medium cost. Clearly, savings can be had if the requirement for expensive supplements needed for the complete growth medium can be reduced. The main question and challenge is how to reduce the media costs in animal cell cultures. Some approaches include:

Use of antibiotics.

Many operators are finding that the use of antibiotics can be dispensed with by providing staff with proper training in aseptic techniques and ensure proper facility & equipment design to minimize contamination. The elimination of antibiotics in animal cell cultures has several advantages:

— *Cost savings.* Antibiotics are expensive.

— *Improve product purification.* Some antibiotics may interfere with isolation and purification of final product(s). Their removal should facilitate product purification.

— *Ineffectiveness against mycoplasma or viral contamination.* The major problem with contamination of animal cell cultures appears to be with mycoplasma; viral infections are less of a problem overall. Antibiotics are generally ineffective against these types of contamination.

— *Eliminate breeding for antibiotic-resistant bacterial strains.* The elimination of antibiotics in cell media will reduce the problem of having to deal with antibiotic-resistant strains.

Use of serum components.

Serum is a complex mixture. It serves a variety of functions. Some of the components present in serum include:

— hormones (for growth stimulation & modulation).

— growth factors (such as epidermal growth factor, fibroblast growth factor, platelet-derived growth factor).

— attachment and spreading factors (eg. fibronectin)

— binding proteins (eg. transferrin).

— nutrients (eg. amino acids - especially glutamine, lipids - VLDL, HDL, LDL, cholesterol, prostaglandins & trace minerals - Fe, Zn, Cu, Mn, Co).

— buffering (eg. albumin) and redox control agents (eg. glutathione).

— protease inhibitor.

— viscosity agent (eg. albumin).

Considering all the diverse & important functions being served by the serum components, it is not surprising that it is difficult to dispense with the use of serum in animal cell cultures. Nevertheless, serum has been investigated extensively in an effort to reduce its use in animal cell cultures. This is because of the following:

— Serum is undefined & may vary from batch to batch.

— The highly nutritious nature of serum is a major source of microbial contamination.

— The presence of serum components can interfere with product isolation and purification.

— Can reduce cost by using less serum. This is often done by optimizing concentration of serum components used. Most of the serum components are provided at the concentrations that are found in normal sera. This means some of the components may be present in much higher concentrations than necessary for optimal growth and productivity of the cell line. Therefore, by optimizing the media with respect to various concentrations of the important components, it may be possible to derive savings. This approach pre-supposes that it's possible to test the effect of each component on productivity.

Of the various factors in serum, the only indispensable items are the growth factors. Because of the availability of these growth factors, it is now possible to cultivate many animal cell lines in a serum-free medium as long as the necessary growth factors are provided. However, the cost of these commercial growth factors, often produced by recombinant cells, may be equal to or greater than that of the serum they replace. Even some serum substitutes are just as costly as serum. At the moment, the decision to use specialized serum-free media is based primarily on productivity considerations rather than saving requirements.

Since different cell lines differ in their serum requirements, it has been suggested that the best way to achieve savings in serum costs in the short run is through more sensible use of serum. For example, serum is routinely used at 10%, but perhaps, 5% is sufficient. In the long run, savings can potentially be made by selecting for cell variants with high productivity in a more defined medium. Alternatively, recombinant DNA methodology may be used to make cells less dependent on growth factors or to produce cheaper growth factors.

Fermentation Considerations

Safety regulations first made in the U.S.A. in 1967 required that biologicals made from animal cells for human consumption must come exclusively from normal diploid anchorage-dependent cell lines. These cells have to be grown on charged surfaces, and cannot be adapted to grow in suspension cultures. It was felt at the time that there was uncertain, unknown and not-easily-appreciated danger(s) associated with using products made by (permanent) cancerous continuous or transformed cells lines. This regulation has affected the development of continuous cell lines, and many commercial operations continue to use normal animal cell lines to produce biological compounds. In the early 1980s, attitudes began to change, and it is now possible to use continuous cell lines to produce some compounds for human consumption, under a rigorous set of conditions.

Anchorage-dependent animal cells

Traditionally, anchorage-dependent monolayer animal cells are maintained in ordinary laboratory glass vessels or culture bottles with some surfaces for attachment. These monolayer cell cultures are often cumbersome and expensive to scale up, require more space than suspension cultures, causes problems in process control in maintenance of a homogeneous pH, oxygen nutrient levels.

Due to the low cell density achievable with conventional Roux bottles or plastic flasks, various methods have been tested to increase the surface area for attachment. A common method is to use roller bottles that can be rotated slowly at 6 to 20 rph (revolutions per hour). Modifications of this method include the "roller bottle with spiral film" or the "roller bottle with glass tubes". Beyond the roller bottles, multitray units and hollow fibre cultures are becoming more common. Some people use the "glass bead" reactor whereby cells grow on the surface of beads and the beads are packed tightly in a column through which medium is continuously perfused by a peristaltic pump.

Another possibility is the microcarrier method for growing anchorage-dependent cells. The microcarriers are typically used in amounts of 3 to 5 mg/mL. It is basically an immobilization method whereby animal cells are attached to beads of DEAE-Sephadex, cross-linked dextran, gelatine, polystyrene, collagen or polyacrylamide. Along the same vein, many animal cells have been immobilized on carriers such as agarose or alginate. The use of microcarriers allow anchorage-dependent cells to be cultivated in a suspension-like manner.

Dense animal cell cultures

Another development is to design systems that will accommodate dense cultures of animal cells for use in production of biologicals. Many of these systems make use of the microcarrier & glass bead techniques or the membrane system. Some of these cultures can achieve cell density of around 5x108 to 109/mL. Cells are packed closely such that they begin to resemble those found in animal tissues. This is like bringing animal cells back to their original organizational state.

Some advantages of dense animal cell cultures include:

— Higher volumetric productivity.

— Smaller fermentation vessels can be used to produce the same product yield.

— Dense cultures often exhibit lower serum requirement per unit cell. This is because each cell occupies a smaller reactor volume.

— Higher product concentration and product yield per volume are possible. This can make it easier for downstream processing to recover the products.

— Potentially, the length of the production runs can be increased. This decreases the cost of producing the cells.

Some disadvantages of dense animal cell cultures include:

— More problems with mass transfer of nutrients.

— Difficult to scale up.

— More problems with process control. This may require more sensitive control devices. With dense cultures, conditions can become unacceptable much quicker if errors in feedback control loop go uncorrected.

Some of the reactor systems used for dense animal cell cultures include:

Air-lift fermenter: To satisfy the requirements of efficient mass transfer and the need to avoid excessive agitation, the air-lift fermenter has been used for the cultivation of many animal cell cultures.

Membrane reactor: Cells are held in a chamber which is separated from fresh media by a membrane that allows for exchanges of gases and nutrients.

The perfusion system: On approach to address some of the problems with dense cultures is the perfusion system. Such a system can be either recirculating or non-recirculating. Cells can be held in place by solid supports or some type of ultrafiltration membranes. In a non-recirculating perfusion system, a fresh medium is being passed through the cells held in a reactor chamber. On exit, the medium is taken away for extraction of the desired product(s). With a recirculating perfusion system, the exiting medium can be reoxygenated (for example through silastic tubing which is permeable to gases such as oxygen, carbon dioxide, etc.).

A filter can be used to trap large particulate debris in the medium before it flows into a reservoir. In the reservoir, additional nutrients can be added; along with mixing & sampling, if necessary. The mixed medium can then be returned to the reactor chamber. These systems functions much like the kidney dialysis machine.

One advantage of the perfusion system, compared to the conventional submerged system, is that some toxic end products can be removed. Another advantage is that the mass transfer problem with dense cultures is reduced to some extent. This is because the medium is fully mixed and oxygenated before returning to the reactor chamber. Also, high cell densities can be achieved.

Technological Uses of Animal Cell Cultures

Animal cells produce specialty high value/low volume compounds that are not readily available by chemical synthesis. Previously, the only available source of these compounds is animal tissue, urine, blood, etc. Nowadays, animal cell cultures may have to compete with production of these compounds by recombinant microorganisms. Some important classes of compounds made by animal cells include:

Viral vaccines

The most important products made by animal cell cultures remain the viral vaccines, which are traditionally made in incubated eggs. For human consumption, only vaccines from normal cells with diploidy and a finite life time can be used. For some vaccines, cell lines are infected during 12th to 13th passage, and production can be maintained for another 40 passages before cells degenerate. Many viral vaccines are produced for human use, the main ones are those against polio, rubella, measles, Hepatitis B, mumps and influenza. Vaccines for veterinary use include those for the Foot and Mouth disease, New Castle's disease, Marek's disease, etc.

Monoclonal antibodies

Monoclonal antibodies are traditionally produced by the hybrid between an antibody-producing spleen cell and a myeloma cell. They are second only to viral vaccines in market value. The market for monoclonal antibodies will continue to increase. They are used for a variety of purposes such as therapeutic agents, research, affinity chromatography, cell typing, diagnostics, immunoassays, etc. A recent development is the production of monoclonal antibodies using recombinant bacteria. This system, based on the use of phage display, has the potential to replace hybridomas in the production of monoclonal antibodies.

Non-antibody immunoregulators

These include interferons, lymphokines and interleukin 2 (formerly T-cell growth factor).

Growth factors

These include epidermal growth factor (EGF), platelet-derived growth factor (PDGF) and fibroblast growth factor (FGF). These compounds have growth stimulating and modulating activity.

Hormones: These include human chorionic gonadotropin, calcitonin, erythropoetin, etc. Presently, human growth hormone & insulin are produced mainly by recombinant microorganisms.

Enzymes: The major one is urokinase (plasminogen activator) for treating thrombosis and embolism.

Toxicity testing: Animal cells can be grown for use in bioassays to assess toxicity of compounds. This aspect is gaining importance as there is a growing trend to minimize the use of live animals in research. Animal cells can serve as the model system for these tests.

2

Enzyme Technology

Enzyme technology is presently going through a phase of maturation and evolution. The maturation is shown by the development of the theory concerning how enzymes function and how this is related to their primary structure through the formation and configuration of their three-dimension structure.

There are many directions in which enzyme technologists are currently applying their art and which are at the forefront of biotechnological research and development. At present, relatively few enzymes are available on a large scale (i.e. > kg) and are suitable for industrial applications. These shortcomings are being addressed in a number of ways:

1. new enzymes are being sought in the natural environment and by strain selection;
2. established industrial enzymes are being used in as wide a variety of ways as can be conceived;
3. novel enzymes are being designed and produce by genetic engineering;
4. new organic catalysts are being designed and synthesised using the 'knowhow' established from enzymology; and
5. more complex enzyme systems are being utilised.

Each of these areas has a extensive and rapidly expanding literature. Some advances possibly belong more properly to other areas of science. Thus, the development of genetically improved enzymes is generally undertaken by molecular biologists and the design and synthesis of novel enzyme-like catalysts is in the provenance of the organic chemists. Both groups of workers will, however, base their science on data provided by the enzyme technologist.

Unnatural Substrates

Many enzymes are not totally specific for their natural substrates. Some have been found to catalyse reactions quite different from those given as the normal reactions and reflected in their name and EC number. Sometimes it necessary to place the enzyme in an unusual environment in order to display new activities. Thus, lipases act as transesterases in primarily non-aqueous environments. In other cases, changes in the environment are not necessary.

Glucose oxidaseb is specific for its reducing substrate (D-glucose) but fairly non-specific in its choice of oxidant, normally molecular oxygen. It has been established that benzoquinone is also an effective electron-accepting substrate:

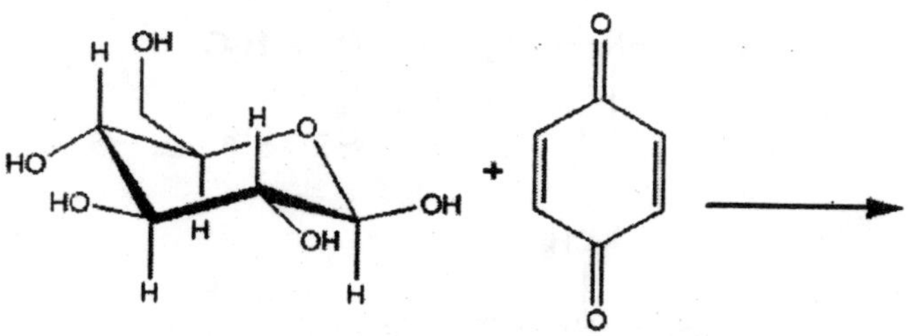

Contd. . .

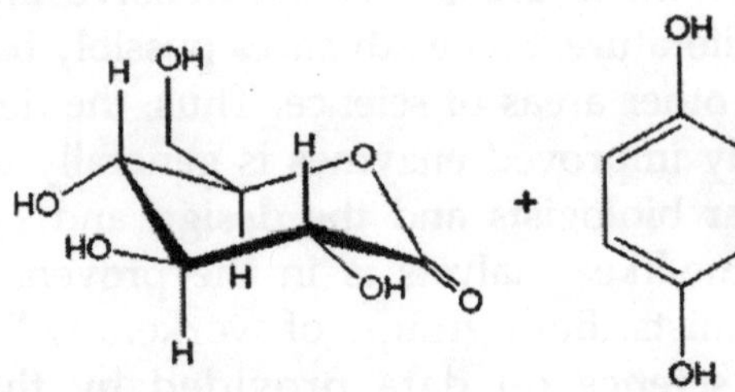

Figure 1. β-D-glucose + benzoquinone ⟶ D-glucono-1,5-lactone + hydroquinone

The product of this reaction, hydroquinone, is a valuable organic chemical, being used in the photographic industry and as an antioxidant. The reaction gives nearly 100% yields, with no possibility of the peroxide-induced inactivation which occurs using molecular oxygen as oxidant. Because of the ready solubility of benzoquinone and low solubility of molecular oxygen, the above reaction (figure 1) can give productivity rates several times greater than the 'natural' reaction.

Acetylcholinesterase normally catalyses the hydrolysis of acetylcholine, the excitatory neurotransmitter, in the synaptic junctions of vertebrates.

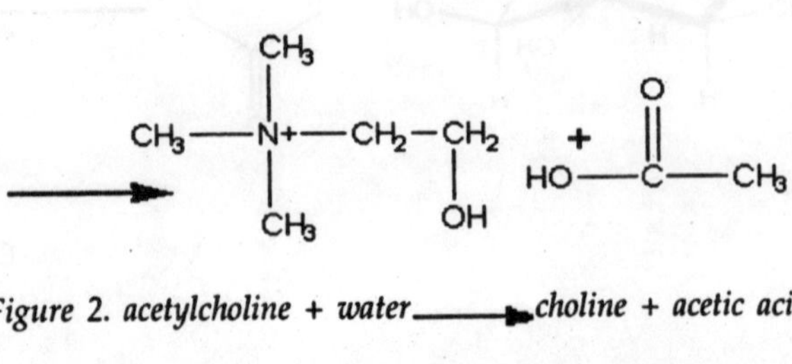

Figure 2. acetylcholine + water ⟶ choline + acetic acid

The acetylcholinesterase from the electric eel has been found additionally to catalyse the stereospecific hydrolysis of acetyl-D-carnitine but not acetyl-L-carnitine.

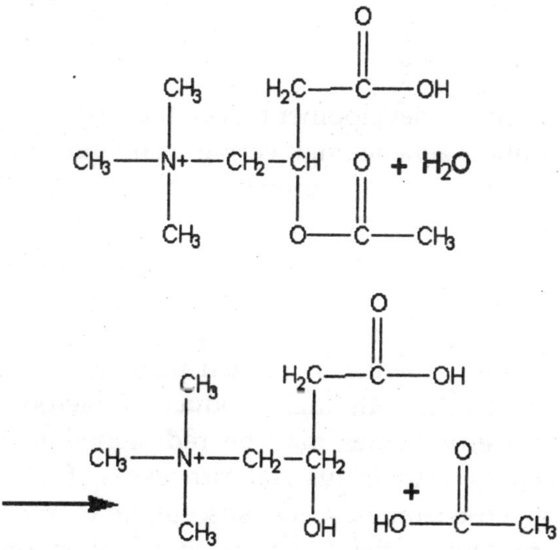

Figure 3. acetyl-D-carnitine + water ——→ D-Carnitine + acetic acid

Although the reaction involving acetyl-D-carnitine has a second-order rate (specificity) constant four orders of magnitude smaller than that utilising acetylcholine, the productivities at high substrate concentrations are comparable, the 'natural' substrate, acetylcholine, causing pronounced substrate inhibition which is not apparent with acetyl-D-carnitine. The 'unnatural' reaction is useful as it may be used in the preparation of L-carnitine, one of the vitamins that has numerous therapeutic applications; the D-isomer being biologically inactive. The enzyme can, therefore, be used in a manner similar to the way in which aminoacylase is used to resolve racemic amino acids.

Chemically synthesised racemic DL-carnitine may be acetylated, using acetyl chloride, to give acetyl-DL-carnitine. The acetylcholinesterase may then be used to produce a mixture of acetyl-L-carnitine and D-carnitine that may be simply resolved by ion-exchange chromatography. The acetyl-L-carnitine is as biologically active as L-carnitine and may be used directly.

Enzyme Engineering

A most exciting development over the last few years is the application genetic engineering techniques to enzyme technology. There are a number of properties which may be improved or altered by genetic engineering including the yield and kinetics of the enzyme, the ease of downstream processing and various safety aspects. Enzymes from dangerous or unapproved microorganisms and from slow growing or limited plant or animal tissue may be cloned into safe high-production microorganisms. In the future, enzymes may be redesigned to fit more appropriately into industrial processes; for example, making glucose isomerase less susceptible to inhibition by the Ca^{2+} present in the starch saccharification processing stream.

The amount of enzyme produced by a microorganism may be increased by increasing the number of gene copies that code for it. This principle has been used to increase the activity of penicillin-G-amidase in Escherichia coli. The cellular DNA from a producing strain is selectively cleaved by the restriction endonuclease HindIII. This hydrolyses the DNA at relatively rare sites containing the 5'-AAGCTT-3' base sequence to give identical 'staggered' ends.

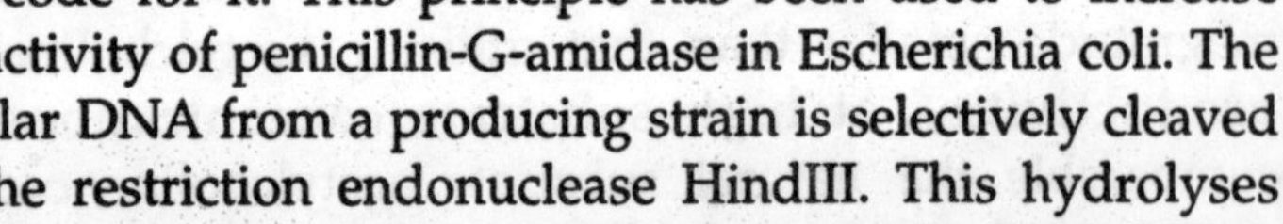

Contd. . .

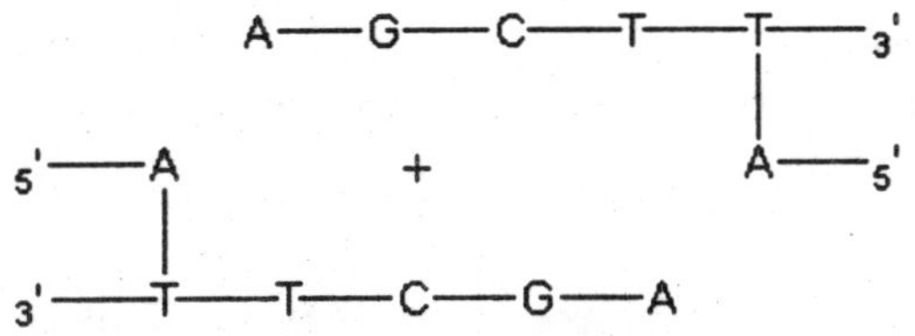

Figure 4. intact DNA ⟶ cleaved DNA

The total DNA is cleaved into about 10000 fragments, only one of which contains the required genetic information. These fragments are individual cloned into a cosmid vector and thereby returned to E. coli. These colonies containing the active gene are identified by their inhibition of a 6-amino-penicillanic acid-sensitive organism. Such colonies are isolated and the penicillin-G-amidase gene transferred on to pBR322 plasmids and recloned back into E. coli.

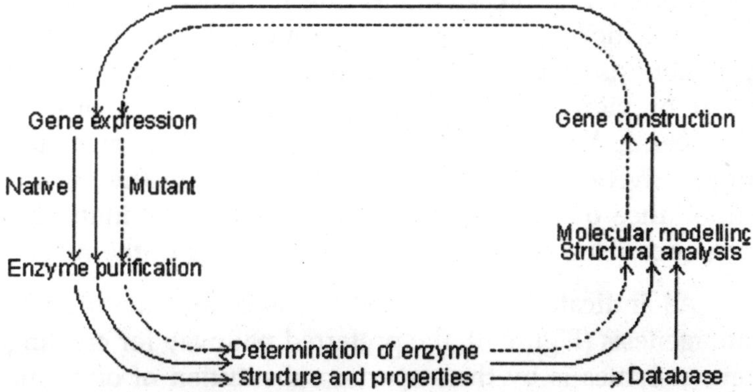

Figure 5. The protein engineering cycle.

The engineered cells, aided by the plasmid amplification at around 50 copies per cell, produce penicillin-G-amidase constitutively and in considerably higher quantities than does the fully induced parental strain. Such increased yields are economically relevant not just for the increased volumetric productivity but also because of reduced downstream processing costs, the resulting crude enzyme being that much purer.

The process starts with the isolation and characterisation of the required enzyme. This information is analysed together with the database of known and putative structural effects of amino acid substitutions to produce a possible improved structure. This factitious enzyme is constructed by site-directed mutagenesis, isolated and characterised. The results, successful or unsuccessful, are added to the database, and the process repeated until the required result is obtained.

Another extremely promising area of genetic engineering is protein engineering. New enzyme structures may be designed and produced in order to improve on existing enzymes or create new activities. An outline of the process of protein engineering is shown in Figure 5. Such factitious enzymes are produced by site-directed mutagenesis (Figure 6). Unfortunately from a practical point of view, much of the research effort in protein engineering has gone into studies concerning the structure and activity of enzymes chosen for their theoretical importance or ease of preparation rather than industrial relevance. This emphasis is likely to change in the future.

As indicated by the method used for site-directed mutagenesis (Figure 6), the preferred pathway for creating new enzymes is by the stepwise substitution of only one or two amino acid residues out of the total protein structure. Although a large database of sequence-structure correlations is available, and growing rapidly together with the necessary software, it is presently insufficient accurately to predict three-dimensional changes as a result of such substitutions. The main problem is assessing the long-range effects, including solvent interactions, on the new structure. As the many reported results would attest, the science is at a stage where it can explain the structural consequences of amino acid substitutions after they have been determined but cannot accurately predict them.

Protein engineering, therefore, is presently rather a hit or miss process which may be used with only little realistic likelihood of immediate success. Apparently quite small sequence changes may give rise to large conformational alterations and even affect the rate-determining step in the enzymic catalysis. However it is reasonable to suppose that, given a sufficiently detailed database plus suitable software, the relative probability of success will increase over the coming years and the products of protein engineering will make a major impact on enzyme technology.

Much protein engineering has been directed at subtilisin (from Bacillus amyloliquefaciens), the principal enzyme in the detergent enzyme preparation, Alcalase. This has been aimed at the improvement of its activity in detergents by stabilising it at even higher temperatures, pH and oxidant strength. Most of the attempted improvements have concerned alterations to:

1. the P1 cleft, which holds the amino acid on the carbonyl side of the targeted peptide bond;
2. the oxyanion hole (principally Asn155), which stabilises the tetrahedral intermediate;
3. the neighbourhood of the catalytic histidyl residue (His64), which has a general base role; and
4. the methionine residue (Met222) which causes subtilisin's lability to oxidation.

It has been found that the effect of a substitution in the P1 cleft on the relative specific activity between substrates may be fairly accurately predicted even though predictions of the absolute effects of such changes are less successful.

Many substitutions, particularly for the glycine residue at the bottom of the P1 cleft (Gly166), have been found to increase the specificity of the enzyme for particular peptide links whilst reducing it for others. These effects are achieved mainly by corresponding changes in the Km rather than the Vmax.

Increases in relative specificity may be useful for some applications. They should not be thought of as the usual result of engineering enzymes, however, as native subtilisin is unusual in being fairly non-specific in its actions, possessing a large hydrophobic binding site which may be made more specific relatively easily (e.g. by reducing its size). The inactivation of subtilisin in bleaching solutions coincides with the conversion of Met222 to its sulfoxide, the consequential increase in volume occluding the oxyanion hole. Substitution of this methionine by serine or alanine produces mutants that are relatively stable, although possessing somewhat reduced activity.

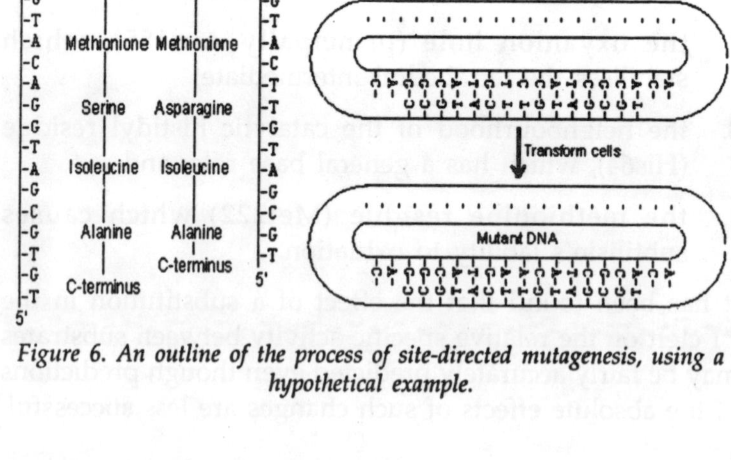

Figure 6. An outline of the process of site-directed mutagenesis, using a hypothetical example.

In figure 6 (a) The primary structure of the enzyme is derived from the DNA sequence. A putative enzyme primary structure is proposed with an asparagine residue replacing the serine present in the native enzyme. A short piece of DNA (the primer), complementary to a section of the gene apart from the base mismatch, is synthesised. (b) The oligonucleotide primer is annealed to a single-stranded copy of the gene and is extended with enzymes and nucleotide triphosphates to give a double-stranded gene. On reproduction, the gene gives rise to both mutant and wild-type clones. The mutant DNA may be identified by hybridisation with radioactively labelled oligonucleotides of complementary structure.

An example of the unpredictable nature of protein engineering is given by trypsin, which has an active site closely related to that of subtilisin. Substitution of the negatively charged aspartic acid residue at the bottom of its P1 cleft (Asp189), which is used for binding the basic side-chains of lysine or arginine, by positively charged lysine gives the predictable result of abolishing the activity against its normal substrates but unpredictably also gives no activity against substrates where these basic residues are replaced by aspartic acid or glutamic acid.

Considerable effort has been spent on engineering more thermophilic enzymes. It has been found that thermophilic enzymes are generally only 20-30 kJ more stable than their mesophilic counterparts. This may be achieved by the addition of just a few extra hydrogen bonds, an internal salt link or extra internal hydrophobic residues, giving a slightly more hydrophobic core. All of these changes are small enough to be achieved by protein engineering. To ensure a more predictable outcome, the secondary structure of the enzyme must be conserved and this generally restricts changes in the exterior surface of the enzyme.

Suitable for exterior substitutions for increasing thermostability have been found to be aspartate → glutamate, lysine → glutamine, valine → threonine, serine → asparagine, isoleucine → threonine, asparagine → aspartate and lysine → arginine. Such substitutions have a fair probability of success. Where allowable, small increases in the interior hydrophobicity for example by substituting interior glycine or serine residues by alanine may also increase the thermostability. It should be recognised that making an enzyme more thermostable reduces its overall flexibility and, hence, it is probable that the factitious enzyme produced will have reduced catalytic efficiency.

Artificial Enzymes

A number of possibilities now exist for the construction of artificial enzymes. These are generally synthetic polymers or oligomers with enzyme-like activities, often called synzymes. They must possess two structural entities, a substrate-binding site and a catalytically effective site. It has been found that producing the facility for substrate binding is relatively straightforward but catalytic sites are somewhat more difficult. Both sites may be designed separately but it appears that, if the synzyme has a binding site for the reaction transition state, this often achieves both functions. For a one-substrate reaction the reaction sequence is given by

$$\text{synzyme} + \text{S} \rightleftharpoons (\text{synzyme-S} \longrightarrow \text{complex})$$
$$\text{synzyme} + \text{P}$$

Figure 7.

Some synzymes are simply derivatised proteins, although covalently immobilised enzymes are not considered here. An example is the derivatisation of myoglobin, the oxygen carrier in muscle, by attaching $(Ru(NH_3)_5)^{3+}$ to three surface histidine residues. This converts it from an oxygen carrier to an oxidase, oxidising ascorbic acid whilst reducing molecular oxygen. The synzyme is almost as effective as natural ascorbate oxidases.

It is impossible to design protein synzymes from scratch with any probability of success, as their conformations are not presently predictable from their primary structure. Such proteins will also show the drawbacks of natural enzymes, being sensitive to denaturation, oxidation and hydrolysis. For example, polylysine binds anionic dyes but only 10% as strongly as the natural binding protein, serum albumin, in spite of the many charges and apolar side-chains.

Polyglutamic acid, however, shows synzymic properties. It acts as an esterase in much thc same fashion as the acid proteases, showing a bell-shaped pH-activity relationship, with optimum activity at about pH 5.3, and Michaelis-Menten kinetics with a Km of 2 mm and V_{max} of 10^{-4} to 10^{-5} s-1 for the hydrolysis of 4-nitrophenyl acetate. Cyclodextrins (Schardinger dextrins) are naturally occurring toroidal molecules consisting of six, seven, eight, nine or ten a-1, 4-linked D-glucose units joined head-to-tail in a ring (a-, b-, g-, d- and e-cyclodextrins, respectively: they may be synthesised from starch by the cyclomaltodextrin glucanotransferase (EC 2.4.1.19) from Bacillus macerans). They differ in the diameter of their cavities (about 0.5^{-1} nm) but all are about 0.7 nm deep. These form hydrophobic pockets due to the glycosidic oxygen atoms and inwards-facing C-H groups.

All the C-6 hydroxyl groups project to one end and all the C-2 and C-3 hydroxyl groups to the other. Their overall characteristic is hydrophilic, being water soluble, but the presence of their hydrophobic pocket enables them to bind hydrophobic molecules of the appropriate size.

Synzymic cyclodextrins are usually derivatised in order to introduce catalytically relevant groups. Many such derivatives have been examined. For example, a C-6 hydroxyl group of b-cyclodextrin was covalently derivatised by an activated pyridoxal coenzyme. The resulting synzyme not only acted a transaminase but also showed stereoselectivity for the L-amino acids. It was not as active as natural transaminases, however.

Polyethyleneimine is formed by polymerising ethyleneimine to give a highly branched hydrophilic three-dimensional matrix. About 25% of the resultant amines are primary, 50% secondary and 25% tertiary:

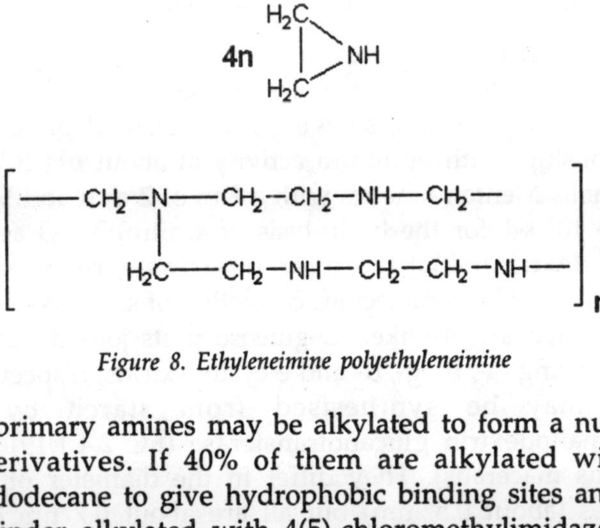

Figure 8. Ethyleneimine polyethyleneimine

The primary amines may be alkylated to form a number of derivatives. If 40% of them are alkylated with 1-iodododecane to give hydrophobic binding sites and the remainder alkylated with 4(5)-chloromethylimidazole to give general acid-base catalytic sites, the resultant synzyme has 27% of the activity of a-chymotrypsin against 4-nitrophenyl esters. As might be expected from its apparently random structure, it has very low esterase specificity. Other synzymes may be created in a similar manner.

Antibodies to transition state analogues of the required reaction may act as synzymes. For example, phosphonate esters of general formula (R-PO2-OR')- are stable analogues of the transition state occurring in carboxylic ester hydrolysis. Monoclonal antibodies raised to immunising protein conjugates covalently attached to these phosphonate esters act as esterases. The specificities of these catalytic antibodies (also called abzymes) depends on the structure of the side-chains (i.e. R and R' in (R-PO2-OR')-) of the antigens.

The Km values may be quite low, often in the micromolar region, whereas the Vmax values are low (below 1 s^{-1}), although still 1000-fold higher than hydrolysis by background hydroxyl ions. A similar strategy may be used to produce synzymes by molecular 'imprinting' of polymers, using the presence of transition state analogues to shape polymerising resins or inactive non-enzymic protein during heat denaturation.

Coenzyme-regenerating Systems

Many oxidoreductases and all ligases utilise coenzymes (e.g. NAD^+, $NADP^+$, NADH, NADPH, ATP), which must be regenerated as each product molecule is formed. Although these represent many of the most useful biological catalysts, their application is presently severely limited by the high cost of the coenzymes and difficulties with their regeneration. These two problems may both be overcome at the same time if the coenzyme is immobilised, together with the enzyme, and regenerated in situ.

A simple way of immobilising/regenerating coenzymes would be to use whole-cell systems and these are, of course, in widespread use. However as outlined earlier, these are of generally lower efficiency and flexibility than immobilised-enzyme systems. Membrane reactors may be used to immobilise the coenzymes but the pore size must be smaller than the coenzyme diameter, which is extremely restrictive. Coenzymes usually must be derivatised for adequate immobilisation and regeneration.

When successfully applied, this process activates the coenzymes for attachment to the immobilisation support but does not interfere with its biological function. The most widely applied synthetic routes involve the alkylation of the exocyclic N6-amino nitrogen of the adenine moiety present in the coenzymes NAD^+, $NADP^+$, NADH, NADPH, ATP and coenzyme A.

In some applications, such as those using membrane reactors it is only necessary that the coenzyme has sufficient size to be retained within the system. High molecular weight water-soluble derivatives are most useful as they cause less diffusional resistance than insoluble coenzyme matrices. Dextrans, polyethyleneimine and polyethylene glycols are widely used. Relatively low levels of coenzyme attachment are generally sought in order to allow greater freedom of movement and avoid possible inhibitory effects.

The kinetic properties of the derived coenzymes vary, depending upon the system, but generally the Michaelis constants are higher and the maximum velocities are lower than with the native coenzymes. Coenzymes immobilised to insoluble supports presently have somewhat less favourable kinetics even when co-immobilised close to the active site of their utilising enzymes. This situation is expected to improve as more information on the protein conformation surrounding the enzymes' active sites becomes available and immobilisation methods become more sophisticated. However, the cost of such derivatives is always likely to remain high and they will only be economically viable for the production of very high value products.

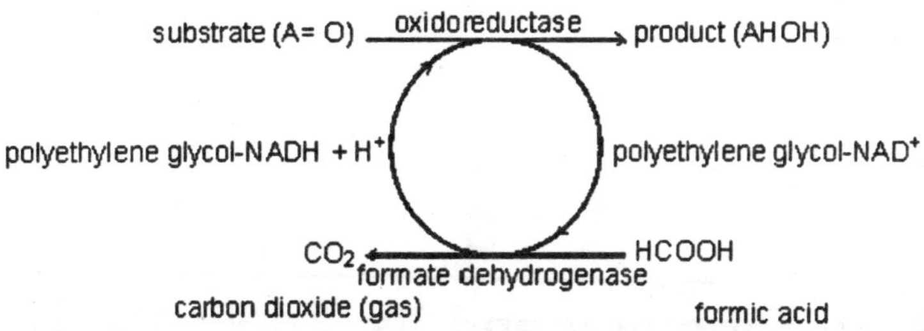

Figure 9.

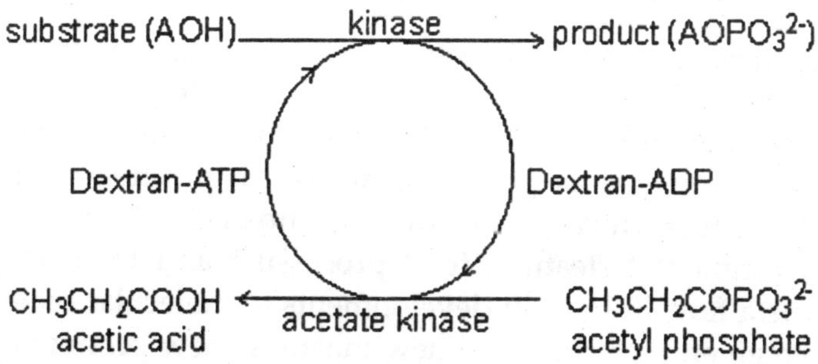

Figure 10.

There are several systems available for the regeneration of the derivatised coenzymes by chemical, electrochemical or enzymic means. Enzymic regeneration is advantageous because of its high specificity but electrochemical procedures for regenerating the oxidoreductase dinucleotides are proving competitive. To be useful in regenerating coenzymes, enzymic processes must utilise cheap substrates and readily available enzymes and give non-interfering and easily separated products. Formate dehydrogenase and acetate kinase present useful examples of their use, although the presently available commercial enzyme preparations are of low activity:

3

Biotechnology and its Applications

The applications of biotechnology are so broad, and the advantages so compelling, that virtually every industry is using this technology. Developments are underway in areas as diverse as pharmaceuticals, diagnostics, textiles, aquaculture, forestry, chemicals, household products, environmental cleanup, food processing and forensics to name a few. But recent developments in molecular biology have given biotechnology new meaning, new prominence, and new potential. It is (modern) biotechnology that has captured the attention of the public.

Modern biotechnology can have a dramatic effect on the world economy and society. One example of modern biotechnology is genetic engineering. Genetic engineering is the process of transferring individual genes between organisms or modifying the genes in an organism to remove or add a desired trait or characteristic. Through genetic engineering, genetically modified crops or organisms are formed. These GM crops or GMOs are used to produce biotech-derived foods.

It is this specific type of modern biotechnology, genetic engineering, that seems to generate the most attention and concern by consumers and consumer groups. What is interesting is that modern biotechnology is far more precise than traditional forms of biotechnology and so is viewed by some as being far safer.)

All organisms are made up of cells that are programmed by the same basic genetic material, called DNA (deoxyribonucleic acid). Each unit of DNA is made up of a combination of the following nucleotides-adenine (A), guanine (G), thymine (T), and cytosine (D)—as well as a sugar and a phosphate. These nucleotides pair up into strands that twist together into a spiral structure call a "double helix." This double helix is DNA. Segments of the DNA tell individual cells how to produce specific proteins. These segments are genes.

More than 10,000 different genes are found in most plant and animal species. This total set of genes for an organism is organized into chromosomes within the cell nucleus. The process by which a multicellular organism develops from a single cell through an embryo stage into an adult is ultimately controlled by the genetic information of the cell, as well as interaction of genes and gene products with environmental factors. When cells reproduce, the DNA strands of the double helix separate. Because nucleotide A always pairs with T and G always pairs with C, each DNA strand serves as a precise blueprint for a specific protein.

Except for mutations or mistakes in the replication process, a single cell is equipped with the information to replicate into millions of identical cells. Because all organisms are made up of the same type of genetic material (nucleotides A, T, G, and C), biotechnologists use enzymes to cut and remove DNA segments from one organism and recombine it with DNA in another organism. This is called recombinant DNA (rDNA) technology, and it is one of the basic tools of modern biotechnology.

rDNA technology is the laboratory manipulation of DNA in which DNA, or fragments of DNA from different sources, are cut and recombined using enzymes. This recombinant DNA is then inserted into a living organism. rDNA technology is usually used synonymously with genetic engineering. rDNA technology allows researchers to move genetic information between unrelated organisms to produce desired products or characteristics or to eliminate undesirable characteristics.

Genetic engineering is the technique of removing, modifying or adding genes to a DNA molecule in order to change the information it contains. By changing this information, genetic engineering changes the type or amount of proteins an organism is capable of producing. Genetic engineering is used in the production of drugs, human gene therapy, and the development of improved plants. For example, an "insect protection" gene (Bt) has been inserted into several crops - corn, cotton, and potatoes - to give farmers new tools for integrated pest management. Bt corn is resistant to European corn borer. This inherent resistance thus reduces a farmers pesticide use for controlling European corn borer, and in turn requires less chemicals and potentially provides higher yielding Agricultural Biotechnology.

Although major genetic improvements have been made in crops, progress in conventional breeding programs has been slow. In fact, most crops grown in the US produce less than their full genetic potential. These shortfalls in yield are due to the inability of crops to tolerate or adapt to environmental stresses, pests, and diseases. For example, some of the world's highest yields of potatoes are in Idaho under irrigation, but in 1993 both quality and yield were severely reduced because of cold, wet weather and widespread frost damage during June.

Some of the world's best bread wheats and malting barleys are produced in the north-central states, but in 1993 the disease Fusarium caused an estimated $1 billion in damage. Scientists have the ability to insert genes that give biological defense against diseases and insects, thus reducing the need for chemical pesticides, and they will soon be able to convey genetic traits that enable crops to better withstand harsh conditions, such as drought (8).

The International Laboratory for Tropical Agricultural Biotechnology (ILTAB) is developing transformation techniques and applications for control of diseases caused by plant viruses in tropical plants such as rice, cassava and tomato. In 1995, ILTAB reported the first transfer through biotechnology of a resistance gene from a wild species of rice to a susceptible cultivated rice variety. The transferred gene expressed resistance to Xanthomonas oryzae, a bacterium which can destroy the crop through disease. The resistant gene was transferred into susceptible rice varieties that are cultivated on more than 24 million hectares around the world.

Benefits can also be seen in the environment, where insect-protected biotech crops reduce the need for chemical pesticide use. Insect-protected crops allow for less potential exposure of farmers and groundwater to chemical residues, while providing farmers with season-long control. Also by reducing the need for pest control, impacts and resources spent on the land are less, thereby preserving the topsoil.

Major advances also have been made through conventional breeding and selection of livestock, but significant gains can still be made by using biotechnology. Currently, farmers in the U.S spend $17 billion dollars on animal health. Diseases such as hog cholera and pests such as screwworm have been eradicated. Uses of biotechnology in animal production include development of vaccines to protect animals from disease, production of several calves from one embryo (cloning), increase of animal growth rate, and rapid disease detection.

Modern biotechnology has offered opportunities to produce more nutritious and better tasting foods, higher crop yields and plants that are naturally protected from disease and insects. Modern biotechnology allows for the transfer of only one or a few desirable genes, thereby permitting scientists to develop crops with specific beneficial traits and reduce undesirable traits.

Traditional biotechnology such as cross-pollination in corn produces numerous, non-selective changes. Genetic modifications have produced fruits that can ripen on the vine for better taste, yet have longer shelf lives through delayed pectin degradation. Tomatoes and other produce containing increased levels of certain nutrients, such as vitamin C, vitamin E, and or beta carotene, and help protect against the risk of chronic diseases, such as some cancers and heart disease.

Similarly introducing genes that increase available iron levels in rice three-fold is a potential remedy for iron deficiency, a condition that effects more than two billion people and causes anemia in about half that number. Most of the today's hard cheese products are made with a biotech enzyme called chymosin. This is produced by genetically engineered bacteria which is considered more purer and plentiful than it's naturally occurring counterpart, rennet, which is derived from calf stomach tissue.

Modern biotechnology offers effective techniques to address food safety concerns. Biotechnical methods may be used to decrease the time necessary to detect foodborne pathogens, toxins, and chemical contaminants, as well as to increase detection sensitivity. Enzymes, antibodies, and microorganisms produced using rDNA techniques are being used to monitor food production and processing systems for quality control. Biotechnology can compress the time frame required to translate fundamental discoveries into applications. This is done by controlling which genes are altered in an organized fashion. For example, a known gene sequence from a corn plant can be altered to improve yield, increase drought tolerance, and produce insect resistance (Bt) in one generation.

Conventional breeding techniques would take several years. Conventional breeding techniques would require that a field of corn is grown and each trait is selected from individual stalks of corn. The ears of corn from selected stalks with each desired trait (e.g, drought tolerance and yield performance) would then be grown and combined (crosspollinated). Their offspring (hybrid) would be further selected for the desired result (a high performing corn with drought tolerance).

With improved technology and knowledge about agricultural organisms, processes, and ecosystems, opportunities will emerge to produce new and improved agricultural products in an environmentally sound manner. In summary, modern biotechnology offers opportunities to improve product quality, nutritional content, and economic benefits. The genetic makeup of plants and animals can be modified by either insertion of new useful genes or removal of unwanted ones. Biotechnology is changing the way plants and animals are grown, boosting their value to growers, processors, and consumers.

Industrial Biotechnology

Industrial biotechnology applies the techniques of modern molecular biology to improve the efficiency and reduce the environmental impacts of industrial processes like textile, paper and pulp, and chemical manufacturing. For example, industrial biotechnology companies develop biocatalysts, such as enzymes, to synthesize chemicals. Enzymes are proteins produced by all organisms. Using biotechnology, the desired enzyme can be manufactured in commercial quantities.

Commodity chemicals (e.g., polymer-grade acrylamide) and specialty chemicals can be produced using biotech applications. Traditional chemical synthesis involves large amounts of energy and often-undesirable products, such as HCl. Using biocatalysts, the same chemicals can be produced more economically and more environmentally friendly. An example would be the substitution of protease in detergents for other cleaning compounds.

Detergent proteases, which remove protein impurities, are essential components of modern detergents. They are used to break down protein, starch, and fatty acids present on items being washed. Protease production results in a biomass that in turn yields a useful byproduct- an organic fertilizer. Biotechnology is also used in the textile industry for the finishing of fabrics and garments. Biotechnology also produces biotech-derived cotton that is warmer, stronger, has improved dye uptake and retention, enhanced absorbency, and wrinkle- and shrink-resistance.

Some agricultural crops, such as corn, can be used in place of petroleum to produce chemicals. The crop's sugar can be fermented to acid, which can be then used as an intermediate to produce other chemical feedstocks for various products. It has been projected that 30% of the world's chemical and fuel needs could be supplied by such renewable resources in the first half of the next century. It has been demonstrated, at test scale, that biopulping reduces the electrical energy required for wood pulping process by 30%.

Environmental Biotechnology

Environmental biotechnology is the used in waste treatment and pollution prevention. Environmental biotechnology can more efficiently clean up many wastes than conventional methods and greatly reduce our dependence on methods for land-based disposal. Every organism ingests nutrients to live and produces by-products as a result. Different organisms need different types of nutrients.

Some bacteria thrive on the chemical components of waste products. Environmental engineers use bioremediation, the broadest application of environmental biotechnology, in two basic ways. They introduce nutrients to stimulate the activity of bacteria already present in the soil at a waste site, or add new bacteria to the soil. The bacteria digest the waste at the site and turn it into harmless byproducts. After the bacteria consume the waste materials, they die off or return to their normal population levels in the environment.

Bioremediation, is an area of increasing interest. Through application of biotechnical methods, enzyme bioreactors are being developed that will pretreat some industrial waste and food waste components and allow their removal through the sewage system rather than through solid waste disposal mechanisms. Waste can also be converted to biofuel to run generators. Microbes can be induced to produce enzymes needed to convert plant and vegetable materials into building blocks for biodegradable plastics.

In some cases, the byproducts of the pollution-fighting microorganisms are themselves useful. For example, methane can be derived from a form of bacteria that degrades sulfur liquor, a waste product of paper manufacturing. This methane can then be used as a fuel or in other industrial processes.

Human Applications

Biotechnical methods are now used to produce many

proteins for pharmaceutical and other specialized purposes. A harmless strain of Escherichia coli bacteria, given a copy of the gene for human insulin, can make insulin. As these genetically modified (GM) bacterial cells age, they produce human insulin, which can be purified and used to treat diabetes in humans. Microorganisms can also be modified to produce digestive enzymes. In the future, these microorganisms could be colonized in the intestinal tract of persons with digestive enzyme insufficiencies.

Products of modern biotechnology include artificial blood vessels from collagen tubes coated with a layer of the anticoagulant heparin. Gene therapy - altering DNA within cells in an organism to treat or cure a disease - is one of the most promising areas of biotechnology research. New genetic therapies are being developed to treat diseases such as cystic fibrosis, AIDS and cancer. DNA fingerprinting is the process of cross matching two strands of DNA.

In criminal investigations, DNA from samples of hair, bodily fluids or skin at a crime scene are compared with those obtained from the suspects. In practice, it has become one of the most powerful and widely known applications of biotechnology today. Another process, polymerase chain reaction (PCR), is also being used to more quickly and accurately identify the presence of infections such as AIDS, Lyme disease and Chlamydia.

Paternity determination is possible because a child's DNA pattern is inherited, half from the mother and half from the father. To establish paternity, DNA fingerprints of the mother, child and the alleged father are compared. The matching sequences of the mother and the child are eliminated from the child's DNA fingerprint; what remains comes from the biological father. These segments are then compared for a match with the DNA fingerprint of the alleged father.

DNA testing is also used on human fossils to determine how closely related fossil samples are from different geographic locations and geologic areas. The results shed light on the history of human evolution and the manner in which human ancestors settled different parts of the world.

Future of Biotechnology Applications

Experts in United States anticipate the world's population in 2050 to be approximately 8.7 billion persons. The world's population is growing, but its surface area is not. Compounding the effects of population growth is the fact that most of the earth's ideal farming land is already being utilized. To avoid damaging environmentally sensitive areas, such as rain forests, we need to increase crop yields for land currently in use. By increasing crop yields, through the use of biotechnology the constant need to clear more land for growing food is reduced.

Countries in Asia, Africa, and elsewhere are grappling with how to continue feeding a growing population. They are also trying to benefit more from their existing resources. Biotechnology holds the key to increasing the yield of staple crops by allowing farmers to reap bigger harvests from currently cultivated land, while preserving the land's ability to support continued farming.

Malnutrition in underdeveloped countries is also being combated with biotechnology. The Rockefeller Foundation is sponsoring research on "golden rice", a crop designed to improve nutrition in the developing world. Rice breeders are using biotechnology to build Vitamin A into the rice. Vitamin A deficiency is a common problem in poor countries. A second phase of the project will increase the iron content in rice to combat anemia, which is widespread problem among women and children in underdeveloped countries. Golden rice, expected to be for sale in Asia in less than five years, will offer dramatic improvements in nutrition and health for millions of people, with little additional costs to consumers.

Similar initiatives using genetic manipulation are aimed at making crops more productive by reducing their dependence on pesticides, fertilizers and irrigation, or by increasing their resistance to plant diseases. Increased crop yield, greater flexibility in growing environments, less use of chemical pesticides and improved nutritional content make agricultural biotechnology, quite literally, the future of the world's food supply.

As biotechnology has become widely used, questions and concerns have also been raised. The most vocal opposition has come from European countries. One of the main areas of concern is the safety of genetically engineered food. In assessing the benefits and risks involved in the use of modern biotechnology, there are a series of issues to be addressed so that informed decisions can be made.

In making value judgments about risks and benefits in the use of biotechnology, it is important to distinguish between technology-inherent risks and technology-transcending risks. The former includes assessing any risks associated with food safety and the behavior of a biotechnology-based product in the environment. The latter involve the political and social context in which the technology is used, including how these uses may benefit or harm the interests of different groups in society.

The health effects of foods grown from genetically engineered crop depend on the composition of the food itself. Any new product may have either beneficial or occasional harmful effects on human health. For example, a biotech-derived food with a higher content of digestible iron is likely to have a positive effect if consumed by iron-deficient individuals. Alternatively, the transfer of genes from one species to another may also transfer the risk for exposure to allergens. These risks are systematically evaluated by FDA and identified prior to commercialization.

Individuals allergic to certain nuts, for example, need to know if genes conveying this trait are transferred to other foods such as soybeans. Labeling would be required if such crops were available to consumers. Among the potential ecological risks identified are increased weediness, due to cross-pollination from genetically modified crops spreads to other plants in nearby fields. This may allow the spread of traits such as herbicide-resistance to non-target plants that could potentially develop into weeds. This ecological risk is assessed when deciding if a plant with a given trait should be released into a particular environment, and if so, under what conditions.

Other potential ecological risks stem from the use of genetically modified corn and cotton with insecticidal genes from Bacillus thuringiensis (Bt genes). This may lead to the development of resistance to Bt in insect populations exposed to the biotech-derived crop. There also may be risks to non-target species, such as birds and butterflies, from the plants with Bt genes. The monitoring of these effects of new crops in the environment and implementation of effective risk management approaches is an essential component of further research. It is also important to keep all risks in perspective by comparing the products of biotechnology and conventional agriculture.

The reduction of biodiversity would represent a technology-transcending risk. Reduced biological diversity due to destruction of tropical forests, conversion of land to agriculture, overfishing, and the other practices to feed a growing world population is a significant loss far more than any potential loss of biodiversity due to biotech-derived crop varieties. Improved governance and international support are necessary to limit loss of biodiversity.

There is no evidence that genetic transfers between unrelated organisms pose human health concerns that are different from those encountered with any new plant or animal variety. The risks associated with biotechnology are the same as those associated with plants and microbes developed by conventional methods.

Survey research over the past decade shows that biotechnology is not likely to become an important issue for most American consumers. Consumers find biotechnology acceptable when they believe it offers benefits and it is safe. Surveys have consistently found that a majority of American consumers are willing to buy insect-protected food crops developed through biotechnology that use fewer chemical pesticides, as well as more nutritious foods. American consumers also appreciate the role that biotechnology can play in feeding the world. Research shows that European consumers are much less supportive of all biotechnology applications.

Surveys since 1992 show that relatively few U.S. consumers have heard or read much about biotechnology. News about the cloned sheep pushed awareness to 50 percent in March 1997. Surveys in the first three months of 2000 show that awareness has fallen back to just over one-third in the United States. Such trends reflect the fact that most people get their information about biotechnology from the media. Unfortunately, many consumers also do not understand some fundamental principles of biology. European consumer awareness is somewhat higher, but knowledge is still low.

In fact, many researches has shown that many consumers are unaware that all foods are derived from plants or animals that already have been genetically modified through traditional (but imprecise) breeding methods. American consumers look to health professionals and scientific experts for credible information, but place relatively little trust in the activists who oppose biotechnology. Research shows that acceptance increases significantly when American consumers learn that organizations such as the National Academy of Sciences and the U.S. Food and Drug Administration have determined that biotech-derived foods are safe.

In contrast, European consumers express the most trust in those groups that oppose biotechnology. They have much less confidence in government, industry, or even scientists. American culture is more supportive and rewarding of new technology. Europeans tend to view food differently from U.S. consumers. In fact, some Europeans reject all American food products. Europeans also want to protect their small farms to maintain open space and rural employment. Such forces underlie much of the European anxiety about agricultural biotech - especially since it is seen as an "American invention."

Many researchers feel that biotechnology has already provided benefits to consumers. Almost all recognize that foods developed through biotechnology have already been part of consumers' everyday diet. They clearly do not agree with most of the opponents' claims and tend to have almost no trust in such groups. Their main concerns involve lack of consumer acceptance-not the safety of the foods. They express high levels of confidence in the science and the regulatory process. In fact, almost none feel that biotechnology should not be used because of uncertain, potential risks. Most food industry leaders do not feel it is necessary to have special labels on biotech-derived foods. They express concerns that such labels would be perceived as a warning by consumers.

Food industry leaders recognize a major need to educate the public about biotechnology. They look to third parties, such as university and government scientists to provide such leadership. Research shows that consumers will accept biotech foods if they see a benefit to themselves or society and if the price is right. Their responses to foods developed through biotechnology are basically the same as for any other food - taste, nutrition, price, safety and convenience are the major factors that influence our decisions about which foods to eat. How seeds and food ingredients are developed will only be relevant for a relatively small group of concerned, consumers.

The potential for public concerns has led several food companies to change their products to avoid biotech-derived ingredients. For example, Gerber Foods received threats from Greenpeace because they had determined the company was using biotech-derived food ingredients. The company firmly believes that the biotech foods are safe to consume. Gerber agreed to drop some of its existing corn and soybean suppliers in favor of ones that can produce crops that are not genetically altered. It became an issue that is suddenly confronting other food companies.

Biotechnology Policy

National governments and international policy making bodies rely on food scientists and others to develop innovations that will create marketable food products and increase food supplies. Governments also rely on scientific research because they are responsible for setting health and safety standards regarding new developments. International organizations can suggest policy approaches and help develop international treaties that are ratified by national governments.

Economic success in the competitive international market demands that food production become more efficient and profitable. National governments and international organizations support food biotechnology as a means to avoid global food shortages. Many policy making bodies are also trying to balance support of the food biotechnology industry with public calls for their regulation. Such regulations are necessary to protect public health and safety, to promote international trade, conserve natural resources, and account for ethical issues.

The majority of processed foods on the market contain soy or corn ingredients that come from GM plants. To date none have posed a food safety risk. The chief safety concerns are the potential to alter nutrient content or introduce allergens. Federal agencies involved in biotechnology regulation include the U.S. Department of Agriculture (USDA) which evaluates agricultural production processes for all foods; the Food and Drug Administration (FDA), which evaluates whole non-animal foods (seafood), food ingredients, and food additives; and the Environmental Protection Agency (EPA), which evaluates plants with insecticidal properties.

Developers of GM plants and biotech-derived foods are required to consult with FDA prior to the commercialization of the product. This consultation procedure entails a science-based safety assessment of the product that focuses on the protection of the consumer, developer, and the environment. Thus developers, have a strong incentive to cooperate fully with FDA and the other agencies prior to marketing their products.

Labeling Issues

Labeling food derived from GM plants and animals is an important, but complex issue. Some consumers and consumer groups believe they have a right to know whether biotechnology was used to produce food. Others believe labeling is not necessary if foods are essentially equivalent in composition. Food labels are regulated by FDA and, in some cases, by USDA. Regulatory agencies are concerned with ensuring that food labels are both true and not misleading. In accordance with their statutory mandate, the FDA has determined that a food product should be labeled as a product of biotechnology only if it has been changed in some significant way. The U.S. Court of Appeals recently reaffirmed that the FDA's labeling policy is correct.

Based on the input from three public hearings, the FDA has recognized that the most effective way to allow for informed choice will be a system of voluntary labeling for foods not produced through biotechnology. If the demand for "GM-free" food is real, a market will become viable. In this case, meaningful choice can be provided to concerned consumers without imposing costs on or denying benefits to the majority of consumers, who support biotechnology. This may be the only viable and cost-effective option for the food industry (given the costs and difficulties associated with segregation and identity preservation.)

National surveys conducted with American consumers between 1997 and 2000 found about three-quarters of consumers supported this FDA labeling policy. There is evidence from focus groups that US consumers are already overwhelmed by the level of detail on food labels and mainly look for relevant information about nutrition. Recent focus groups have demonstrated that the wording on labels has a significant effect on consumer understanding and acceptance of biotechnology. Any label messages need to be simple and clear. However, labeling will be counterproductive without education. It is important to understand when FDA will and won't require a product label. The same labeling laws that apply to all other foods and food ingredients apply to products of food biotechnology. Requirements for all food labels mandate proper identification of products and notice of health or safety concerns. Labeling would be required in the following instances - but not simply because the products were made using biotechnology.

— Potential food allergy is an example of a health or safety risk that would mandate a product label (e.g., if a gene has been transferred from a known allergen, such as a peanut).

— When substantial changes have been made to the expected nutritional composition of a food (e.g., vitamins have been added to a particular type of produce).

— If the common identity of a food product has been changed (e.g., if the composition of a fruit or vegetable becomes significantly different it would have to have a different name).

Labeling of biotech-derived foods will be required in some countries (such as Japan and the European Union) for cultural and religious reasons. These countries also believe labeling should be required simply because the consumers want to know more about the content of the food or how it was produced. Some have suggested that such labeling initiatives have been driven by economic and political interests rather than by scientific need. Consumers in these countries will ultimately end up paying even higher prices for their food as a result of such strict labeling policies.

4

Biomimetic Concepts to Asymmetric Synthesis

In organic chemistry "value" is directly related to purity; in most instances this implies that an enantiomerically pure product is wanted. In recent years the number of methods available for highyielding and enantioselective transformation of organic compounds has increased tremendously. Most of the newly introduced reactions are catalytic in nature. Clearly, catalytic transformation provides the best "atom economy", because the stoichiometric introduction and removal of (chiral) auxiliaries can be avoided, or at least minimized.

Recently, the catalysts employed for enantioselective synthesis of organic compounds such as pharmaceutical products, agrochemicals, fine chemicals, or synthetic intermediates, fell into two general categories - transition metal complexes and enzymes. In 2001 the Nobel Prize in Chemistry was awarded to William R. Knowles and Ryoji Noyori "for their work on chirally catalyzed hydrogenation reactions", and to K. Barry Sharpless "for his work on chirally catalyzed oxidation reactions". Could there be a better illustration of the importance of asymmetric catalysis?

For all three laureates the development of chiral transition metal catalysts was the key to success. It has been a long-standing belief that only man-made transition metal catalysts can be tailored to produce either of two product enantiomers whereas enzymes cannot. This dogma has been challenged in recent years by tremendous advances in the field of biocatalysis, for example the discovery of preparatively useful enzymes from novel organisms, and the optimization of enzyme performance by selective mutation or by evolutionary methods.

The complementary character of both types of catalyst becomes obvious. Between the extremes of transition metal catalysis and enzymatic transformations, a third approach to the catalytic production of enantiomerically pure organic compounds has emerged—organocatalysis. Organocatalysts are purely "organic" molecules, i.e. composed of (mainly) carbon, hydrogen, nitrogen, sulfur and phosphorus. As opposed to organic ligands in transition metal complexes, the catalytic activity of organocatalysts resides in the low-molecular-weight organic molecule itself, and no transition metals (or other metals) are required.

Organocatalysts have several advantages. They are usually robust, inexpensive and readily available, and non-toxic. Because of their inertness toward moisture and oxygen, demanding reaction conditions, for example inert atmosphere, low temperatures, absolute solvents, etc., are, in many instances, not required. Because of the absence of transition metals, organocatalytic methods seem to be especially attractive for the preparation of compounds that do not tolerate metal contamination, e.g. pharmaceutical products. A selection of typical organocatalysts is shown in Figure 1.

Proline, a chiral-pool compound which catalyzes aldol and related reactions by iminium ion or enamine pathways, is a prototypical example. The same is true for cinchona alkaloids such as quinine, which has been

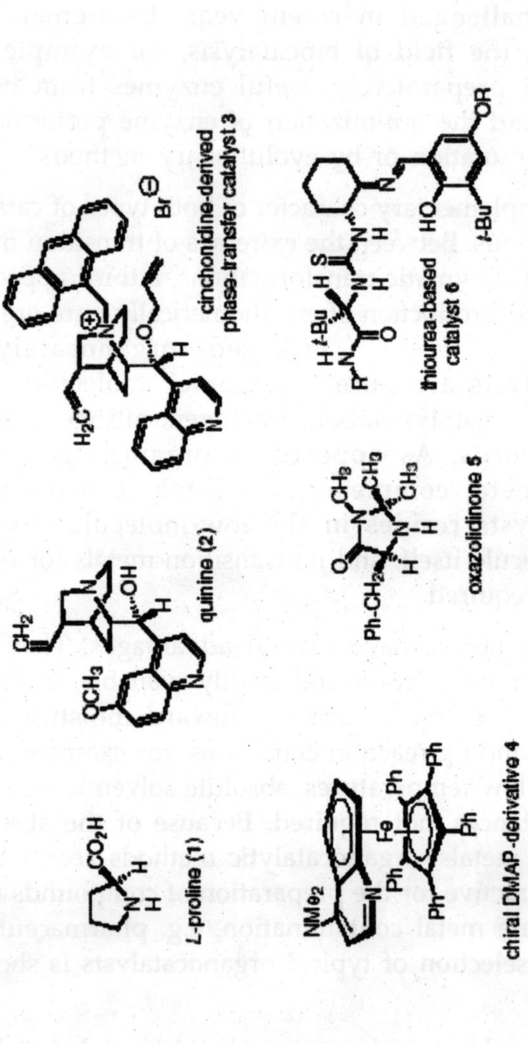

L-proline (1)

quinine (2)

cinchonidine-derived phase-transfer catalyst 3

chiral DMAP-derivative 4

oxazolidinone 5

thiourea-based catalyst 6

Contd. . .

Contd. . .

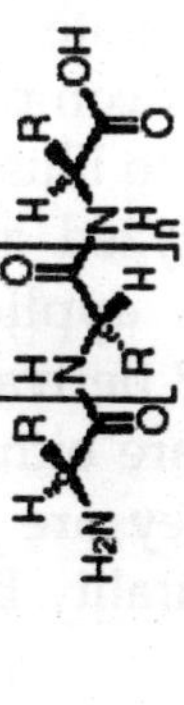

oligo-*L*-Leu 7, R: *iso*-butyl

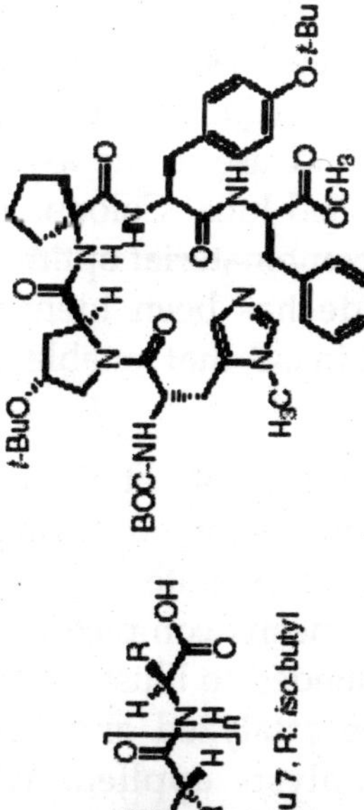

pentapeptide 8

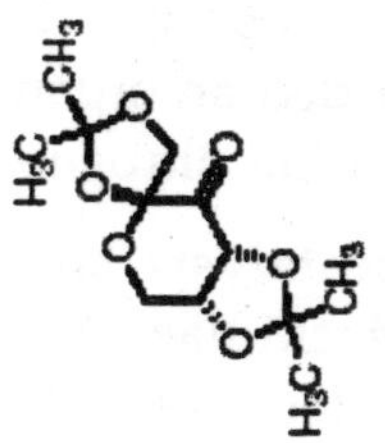

D-fructose-derived ketone 9

Figure 1.

abundantly used as a chiral base or as a chiral nucleophilic catalyst and which has served as the basis for many highly enantioselective phase-transfer catalysts. The latter are exemplified by which enables, e.g., the alkylation of glycine imines with very high enantioselectivity. The planar chiral DMAP derivative introduced by Fu et al. is extremely selective in several nucleophilic catalyses. Although it is a ferrocene it is regarded an organocatalyst because its "active site" is the pyridine nitrogen atom.

Amino acid-derived organocatalysts such as the oxazolidinone 5 introduced by MacMillan et al. or the chiral thiourea 6 introduced by Jacobsen et al. have enabled excellent enantioselectivity in, e.g., Diels-Alder reactions of a,b-unsaturated aldehydes (oxazolidinone 5) or the hydrocyanation of imines (thiourea 6). Peptides, such as oligo-l-leucine have found use in the asymmetric epoxidation of enones, the so-called Julia´-Colonna reaction . Peptides are ideal objects for combinatorial optimization/ selection, and the pentapeptide has been identified by Miller et al. as an artificial kinase that enables highly enantioselective phosphorylation.

The chiral ketone introduced by Shi et al. is derived from d-fructose and catalyzes the asymmetric epoxidation of a wide range of ole.ns with persulfate as the oxygen source. This small (and by no means complete) selection of current organocatalysts is intended to illustrate the wide range of reactions that can be catalyzed and the ready accessibility of the organocatalysts applied. With the exception of the planar chiral DMAP derivative 4, all the organocatalysts shown in Figure 1 are either chiral-pool compounds themselves (1, 2), or they are derived from these readily available sources of chirality by means of a few synthetic steps.

The historic roots of organocatalysis go back to the use of low-molecular-weight compounds in an attempt both to understand and to mimic the catalytic activity and selectivity of enzymes. As early as 1928 the German chemist Wolfgang Langenbeck published on "Analogies in the catalytic action of enzymes and de.nite organic substances".

The first example of an asymmetric organocatalytic reaction was reported by Bredig and Fiske as early as 1912, i.e. ca. 90 years ago. These two German chemists reported that addition of HCN to benzaldehyde is accelerated by the alkaloids quinine and quinidine and that the resulting cyanohydrins are optically active and of opposite chirality.

Unfortunately, the optical yields achieved in most of these early examples were in the rangea10% and thus insu.cient for preparative purposes. Pioneering work by Pracejus et al. in 1960, again using alkaloids as catalysts, a.orded quite remarkable 74% ee in the addition of methanol to phenylmethylketene. In this particular reaction 1 mol% O-acetylquinine (10, Figure 2) served as the catalyst.

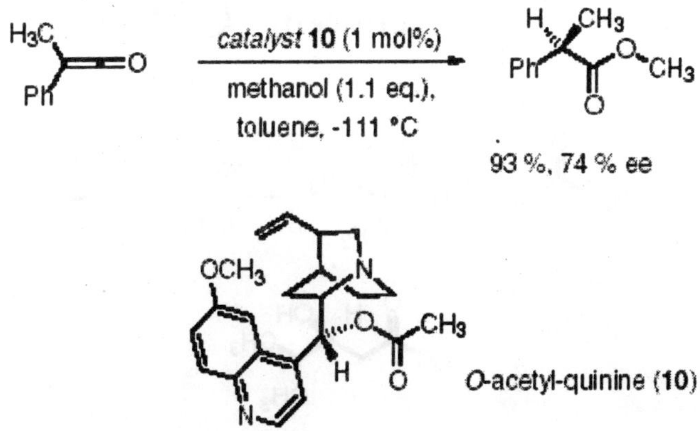

O-acetyl-quinine (10)

Figure 2.

Further breakthroughs in enantioselectivity were achieved in the 1970s and 1980s. For example, 1971 saw the discovery of the Hajos-Parrish-Eder-Sauer-Wiechert reaction, i.e. the proline (1)-catalyzed intramolecular asymmetric aldol cyclodehydration of the achiral trione to the unsaturated Wieland-Miescher ketone (Figure 3). Ketone is an important intermediate in steroid synthesis.

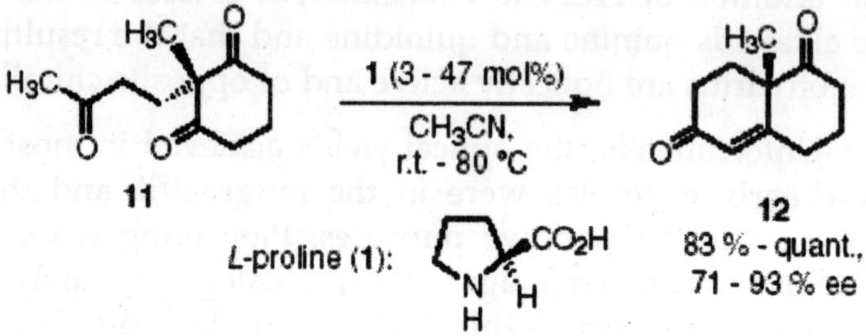

Figure 3.

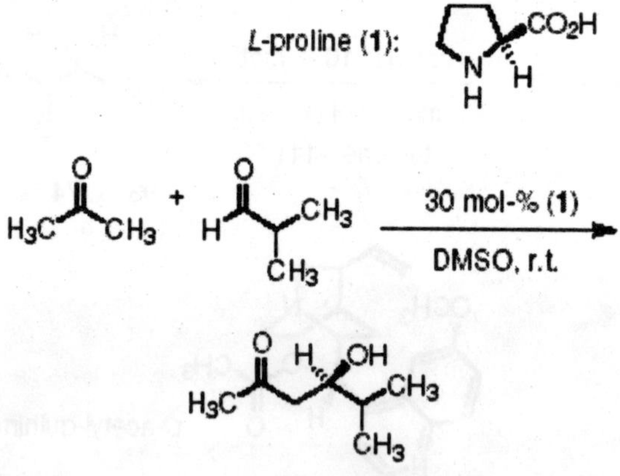

13, 97 %, 96 % ee

Proline (1)-catalyzed intermolecular aldol reaction *Contd..*

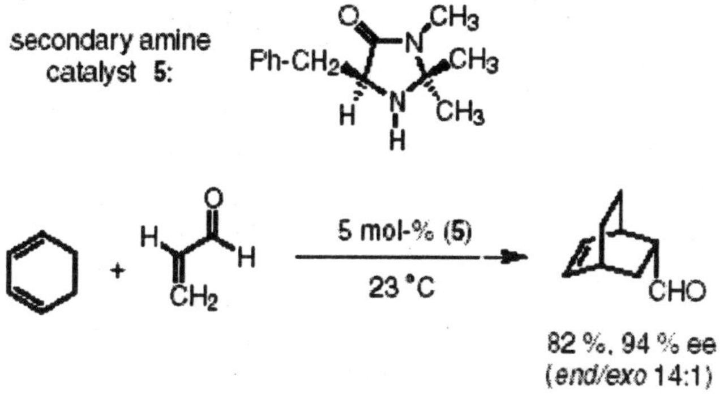

Secondary amine 5-catalyzed Diels-Alder reaction

Figure 4.

Surprisingly, the catalytic potential of proline in asymmetric aldol reactions was not explored further until recently. List et al. reported pioneering studies in 2000 on intermolecular aldol reactions. For example, acetone can be added to a variety of aldehydes, a.ording the corresponding aldols in excellent yields and enantiomeric purity. The example of iso-butyraldehyde as acceptor is shown in Figure 4. In this example, the product aldol 13 was obtained in 97% isolated yield and with 96% ee. The remarkable chemo- and enantioselectivity observed by List et al. triggered massive further research activity in proline-catalyzed aldol, Mannich, Michael, and related reactions. In the same year, MacMillan et al. reported that the phenylalanine-derived secondary amine 5 catalyzes the Diels-

Alder reaction of a,b-unsaturated aldehydes with enantioselectivity up to 94% (Figure 4). MacMillan et al. was followed by numerous further applications of the catalyst and related secondary amines. A similarly remarkable event was the discovery of the cyclic peptide shown in Figure 5. In 1981 this cyclic dipeptide - readily available from l-histidine and l-phenylalanine - was reported, by Inoue et al., to catalyze the addition of HCN to benzaldehyde with up to 90% ee (Figure 5). Again, this observation sparked intensive research in the .eld of peptide-catalyzed addition of nucleophiles to aldehydes and imines.

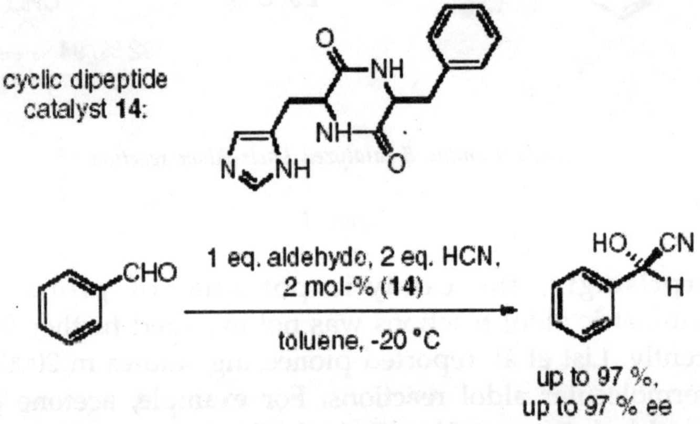

Figure 5. The cyclo-L-*His*-L-*Phe catalyst*

Also striking was the discovery, by Julia´, Colonna et al. in the early 1980s, of the poly-amino acid-catalyzed epoxidation of chalcones by alkaline hydrogen peroxide. In this experimentally most convenient reaction, enantiomeric excesses > 90% are readily achieved (Figure 6).

Asymmetric organocatalysis is, in principle, an "old" branch of organic chemistry, with its beginnings dating back to the early 20th century (for example the .rst asymmetric hydrocyanation of an aldehyde in 1912). This initial phase of organocatalysis was, however, mainly mechanistic/biomimetic in nature, and the relatively low enantiomeric excess achieved prohibited "real" synthetic applications. Isolated examples of highly enantioselective organocatalytic processes were reported in the 1960s to the 1980s, for example the alkaloidcatalyzed addition of alcohols to prochiral ketenes by Pracejus et al. (Figure 2), the Hajos-Parrish-Eder-Sauer-Wiechert reaction (Figure 3), the hydrocyanantion of aldehydes using the Inoue catalyst 14 (Figure 5), or the Julia´-Colonna epoxidation (Figure 6), but the field still remained "sub-critical".

poly-amino acid **15**:

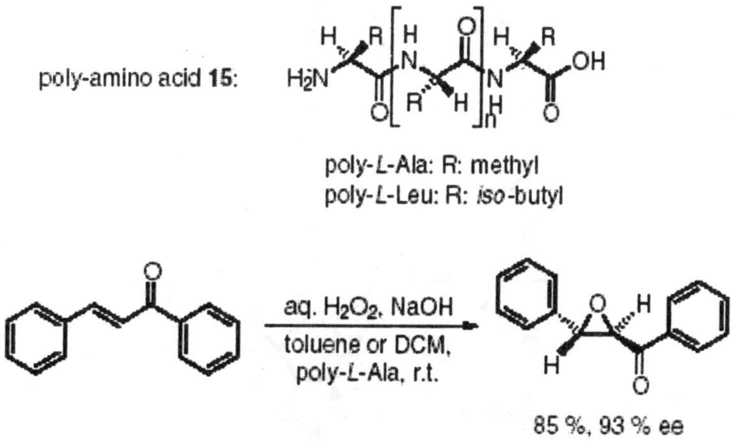

Figure 6.

Now, triggered by the ground-breaking work of List, MacMillan, and others in the early 2000s, the last ca. five years have seen exponential growth of the .eld of asymmetric organocatalysis. Iminium and enamine-based organocatalysis now enables cycloadditions, Michael additions, aldol reactions, nucleophilic substitutions, and many other transformations with excellent enantioselectivity; new generations of phase-transfer catalysts give almost perfect enantiomeric excesses at low catalyst loadings; chiral ureas and thioureas are extremely

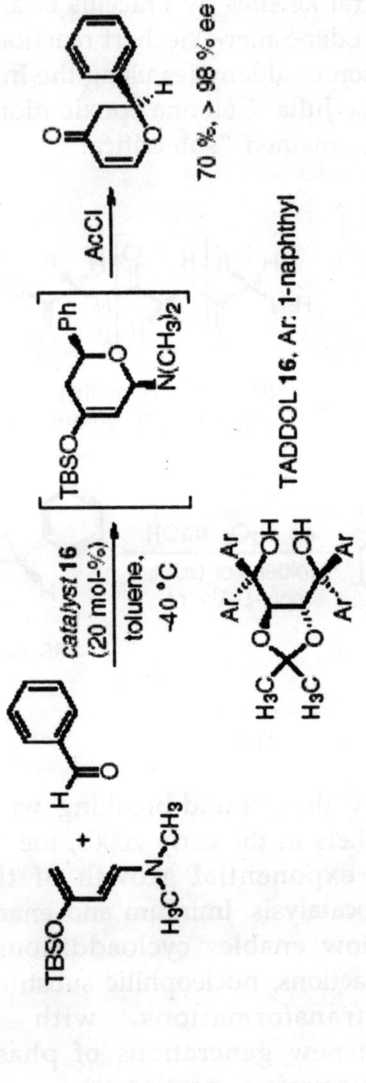
catalyst 16
(20 mol-%)
toluene,
-40 °C
AcCl
70 %, > 98 % ee
TADDOL 16, Ar: 1-naphthyl

Figure 7.

enantioselective catalysts for addition of a variety of nucleophiles to aldehydes and imines; and so forth.

Organocatalysis currently seems to be in the state of a "gold rush" and at short intervals new "gold mines" are discovered and reported in the literature. A very recent example is the finding by Rawal et al. that hetero-Diels-Alder reactions - a classical domain of metal-based Lewis acids - can be effected with very high enantioselectivity by hydrogen bonding to chiral diols such as TADDOL (Figure 7).

Compared with earlier approaches, both prospecting and exploiting of the fields is greatly aided and accelerated by advanced analytical technology and, in particular, by synergism with theoretical and computational chemistry. Overall, asymmetric organocatalysis has matured in recent few years into a very powerful, practical, and broadly applicable third methodological approach in catalytic asymmetric synthesis.

5

Enzymes in Organic Media

The use of enzymes in organic media (with low water content) has been one of the most exciting facets of enzymology in recent times. It is an area in which applications and phenomena preceded the understanding of catalysis at the molecular level. Also, it is an area in which enzymologists continue to discover unexpected and yet fundamental behaviour of biocatalysts. In some cases (e.g.ultrasonoenzymology and microwave-assisted nonaqueous enzymology), this is simply because these aspects are what the media calls breaking news.

While free enzyme powders continue to be used in nearly anhydrous media, a variety of forms and formulations have been described. There have been two major motivations behind evolving these designs. The . rst one has been the perception that enzymes are not stable in neat organic solvents. Unfortunately, a clear distinction between exposure to organic media and working in such media has not been made by many workers.

When enzymes are exposed to such media (stress or denaturing conditions) for a limited time, and are. recovered and checked for biological activity in water, early studies indicate that full activity is observed. This implies that no irreversible denaturation has taken place.

On the other hand, enzymes in different solvents display different k_{cat}/K_m, so operational stability is different in different organic solvents.

A number of solvent parameters have been described; some have raised the question of nonpolar solvents not being the same as hydrophobic solvents. T he most acceptable, even if not totally satisfactory, parameter is log P, where log P is the partition coef. cient of the solvent for the standard octanol/ water two-phase system. Thus, early efforts, and even some recent ones, have aimed at stabilization of enzymes in organic media. Chemical modification, chemical crosslinking, immobilization, protein aggregation and protein engineering have all been tried. Of these, immobilization has been used most often. The most popular immobilization method is, in fact, codrying or deposition rather than immobilization.

The second kind of design has been prompted by the appreciation that kcat/Km in organic media (with low water content) is far less than the corresponding k_{cat}/K_m in aqueous media because drying by lyophilization causes reversible damage to the enzyme structure. Thus, it is necessary to either prevent this damage or use other methods of drying. The use of additives (cryoprotectants, lyoprotectants, salts, etc.) during lyophilization falls in the former category.

The latter approach consists of precipitating enzymes by organic solvents. Some of the forms and formulations based upon both these approaches are: lyophilized powders containing cryoprotectants or/and lyoprotectants, propanol rinsed preparations, enzyme precipitated with and rinsed with propanol, enzymes obtained after three-phase partitioning, crosslinked enzyme crystals, crosslinked enzyme aggregates and enzymes coated with ionic liquid.

Additives and Activation

Lyophilized enzymes are the most common forms of enzymes used in enzymology. Lyophilization involves (a) initial freezing, (b) primary drying (in which ice separated from the protein phase is removed by sublimation) and © secondary drying (in which the bound or trapped water is removed from proteins). Basically, lyophilization (or freeze-drying) involves two denaturing conditions: freezing and drying. Cryo protectants (very heterogeneous as far as chemical structures are concerned (e. g. sugars, amino acids, polyols, salts, etc.) prevent structural damage due to freeze stress. This protection may be due to preferential exclusion.

During the drying stage (dehydration), one needs lyoprotectants. Lyoprotectants substitute water molecules, which are being removed, by forming H-bonds with the protein structure. Sucr ose is a good example. Evidence (most of it based upon FTIR) exists, which shows that the absence of lyoprotectant leads to both the random structure and the a-helical structure being partially converted to b-sheet structure (in which peptide bonds are linked to each other via H-bonds).

In conventional enzymology (in aqueous media), rehydration reverses much of the structural damage during lyophilization. In nonaqueous enzymology, this becomes very crucial and is now considered as largely responsible for the low k_{cat}/K_m of enzymes in nearly anhydrous media. The k_{cat} values for esterase (in water) and transesterification (in hexane) activities of subtilisin have been found to be 5.9×10^3 and 1.04×10^{-1} $M^{-1}Æ^{s-1}$, respectively.

Dabulis & Klibanov showed that addition of a variety of substances (an amino acid derivative, sugars, polyethylene glycol) during lyophilization resulted in more active (in the range of 3-321-fold) preparation of seven hydrolytic enzymes. A few years later, Triantafyllou et al.

carried out a more intensive work on the effect of sorbitol and other additives on α-chymotrypsin and some lipases. While Dabulis & Klibanov found that higher activity persists even if excipients are removed, Triantafyllou et al. found that 'washing to remove sorbitol had a negative effect on the activity' of both lipase and a-chymotrypsin in their immobilized forms. One of the most remarkable results with additives has been the salt activation reported by Khmelnitsky et al.

The 3750 times higher activity of subtilisin in the presence of 98% (w/w) of KCl was tentatively explained by either rigid salt structure protecting the enzyme in organic solvents or polar environment of salt helping in maintaining native structure of the enzyme. Soon after, Bedell et al. provided evidence that it is not a result of reduced diffusional limitation. Subseq uently, it was shown that salt activation is the result of kosmotropicity and this stabilizing effect (during lyophilization) operates via preferential hydration.

Also, a combination of kosmotropic salt sodium acetate and sodium carbonate buffer had a cumulative effect. Hsu & Dordick found that salt activation enhances enantioselectivity of subtilisin because the favoured reaction is more sensitive to the structural integrity of the enzyme. On the other hand, Altreuter et al. found that salt activation is accompanied by expanded and unnatural regioselectivity of subtilisin. The latter feature is very valuable in the use of enzymes for developing combinatorial biocatalysis.

Recently, Lindsay et al. reported the salt activation of a nonprotease, penicillin amidase and found that salt activation was dependent upon the water content in the solvent medium. However, the preparation lyophilized in the presence of trehalose behaved differently. The conclusion was that salt activation is mechanistically distinct from lyoprotection.

Medium Engineering

The activity of enzymes was higher in hydrophobic solvents than in hydrophilic ones. This issue in a comprehensive fashion and pointed out that log P, as a solvent parameter, correlates best with enzyme activity. Narayan& Klibanov pointed out that it is hydrophobicity and not polarity or water miscibility which is important. As this may lead to some confusion among biochemists, it is best to quote 'The log P parameter can be called a measure of solvent 'hydrophobicity', which is an accurate description of what affects its value. This contrasts with other parameters such as dielectric, whichmeasure the bulk polarity. One illustration of the difference is to consider a homologous series of solvents. Adding extra methylene groups to an alcohol, for example, will cause a regular increase in hydrophobicity.

Thus, decanol is more hydrophobic (higher log P) than hexane, but will be more polar by almost any measure of bulk properties. It is also pertinent to add that most often, the medium effects are explained in terms of more hydrophobic solvents having less tendency to strip water from enzymes. Su ch medium effects will not be seen if identical water activities in both media are used. Many solvent effects, in fact, arise from substrate solvation. If a substrate is more soluble in a solvent, the substrate molecules will be less available to the enzyme.

Thus, higher K_m values may lead to lower reaction rates. The nature of the solvent media affects both the enantio- and regioselectivity of the enzymes. Var ious hypotheses have been given to rationalize these important effects. No consensus seems to have emerged so far. Finally, one must mention solvent-free media wherein one or more of the substrate(s) form the medium and no other solvent is required.

In some cases, this results in more ef. cient conversion rates. A good example is the production of biodiesel with lipases. Biodiesel consists of monoalkyl esters of long chain fatty acids. There is no petroleum or other fossil fuel in biodiesel; the diesel part of its name is based upon the fact that it can be substituted in place of petroleum diesel fuel. It is produced from vegetable oils or fats by lipase catalyzed transesteri. cation with methanol or ethanol.

Water activity in Anhydrous Media

It is now recognized that less than a monolayer of water is needed for an enzyme molecule to start showing biological activity. Beyond this, addition of more water molecules increases biological activity. Howe ver, in the case of hydrolases, beyond a threshold limit of water concentration, hydrolytic activity starts effectively competing with transferase or synthetic activity.

Working with alcohol dehydrogenase in a variety of solvents, they determined the adsorption isotherm of water on the enzyme in a variety of solvents. It was observed that the water present on the enzyme correlated well with the reaction rates in different solvents. The higher the water content on the enzyme, the higher was the reaction rate. The rationale which emerged was that when water is added to the reaction system, it partitions among the enzyme, solvent, headspace of the reaction vessel and the immobilization matrix.

Controlled addition of water continues to be used because of its simplicity. The relationship between mole fraction of water and activity in the case of several organic solvents has been reported. Saturated solutions of different salts are known to release and absorb water at constant humidity. Wehtje et al. have described a method by which the water activity can be kept constant as the reaction proceeds. A saturated solution of the appropriate salt is pumped through silicon tubing that passes through the reaction mixture. As the water molecules can diffuse out of the tubing, the system remains equilibrated at the same water activity throughout.

Another method, which is becoming increasingly popular, is the use of a pair of salt hydrates to maintain a constant water activity level throughout the reaction. Here the lower hydrate or anhydrous form will absorb water and the higher hydrate will release it. A few years ago, suitable salt hydrates for different water activity levels were listed. One area which needs more work is the effect of water activity on the stereoselectivity of the reaction. There are already a few reports which show that both rate and enantiomeric ratio change simultaneously when water activity is increased.

pH Tuning and pH Memory

The correct protonation state of side chains of amino acid residues of enzymes is important in nonaqueous media as well. Hence pH tuning (placing the enzyme in water at optimum pH of the enzyme, and lyophilizing) results in higher rates in organic solvents. However, this is valid only if the reaction does not change the acid/base concentration.

Microwave-assisted Reactions in Nonaqueous Enzymology

The microwave region (0. 3-300 GHz) falls between the infrared and radiofrequency regions of the electromagnetic spectrum. Chemical synthesis using microwave irradiation has been extensively reported. In this approach, microwave irradiation replaces conventional forms of heating. Upon irradiation with microwaves, a polar molecule continually aligns itself with the fluctuating field. This converts electromagnetic energy into heat energy. It is not very clear whether these rate enhancements are purely due to thermal effects or whether some nonthermal effects are involved.

That 'the rate of a reaction in many syntheses is so high that it can not be accounted for by heating effects only'. Early work on microwave-assisted enzymatic reactions in water was naturally limited to the use of thermostable enzymes. Considering that enzymes are highly thermostable in nearly anhydrous media, it is natural that microwave assistance is also explored for enzymatic catalysis in such media. Also, the commercial availability of microwave ovens, which are able to maintain constant temperature by remote IR feedback control, has also facilitated work in this area.

Carrillo-Munoz et al. used two commercially available lipases from Pseudomonas cepacia and Candida antarctica for esteri. cation and transesteri. cation reactions at a constant temperature of around 100°C. The enzymes were immobilized on matrices which did not absorb microwaves. The microwave assistance was found to enhance both initial rates as well as enantioselectivity as compared to the results obtained under conditions of classical heating. An interesting result was that the use of substrates with greater polarity led to greater effect by microwaves.

The results of Parker et al., on the other hand, emphasized the importance of optimization of water activity for obtaining maximum increase (two- to threefold) by microwave assistance. The molecularmechanisms for these results are seldom discussed and are not well understood. Chen et al. have provided a protocol for using microwave assistance for protease catalyzed peptide synthesis. Unfort unately, most of these protocols do not take account of temperature control during microwave irradiations. Recently, esterification by α-chymotrypsin and transesteri . cation by subtilisin Carlsberg were accelerated (in the range of 2. 1-4. 7 times) in six solvents of differing polarities and at different water activities.

Interestingly, microwave irradiation could be used in conjunction with pH tuning and salt activation. For example, at the same level of water activity (0. 3%, v/v) in n-octane, untuned subtilisin showed a transesteri. cation rate of 0. 5 mmolÆh)1 at 25°C. The salt-activated and pH tuned subtilisin showed about 20 times increase in reaction rates when microwave irradiations were used at 25°C. It was also observed that the choice of the reaction medium was a factor which dominated the microwave effect.

Maugard et al. have exploited microwave assistance for a somewhat different kind of application, i. e. synthesis of galactooligosaccharides from lactose by using Kluyveromyces lactis β-galactosidase. Low water activity, high lactose concentration and cosolvents with higher log P value all favoured oligosaccharide synthesis. By optimizing conditions and using microwave irradiations, the synthesis of galactooligosaccharides could be increased 217 times. It is hoped that when more experience with different systems becomes available, microwave assistance will become a powerful approach in nonaqueous enzymology.

Ultrasonication

Ultrasound waves have frequencies beyond the normal range of hearing, which consists of 20 kHz to 100 MHz and beyond. High power ultrasound waves can generate cavitation within a liquid. Cavitation provides a source of energy which can facilitate chemical processes. In this case, cavitation is an induced bubble activity. Cavitation can generate very high local pressure (up to 1000 atm) and very high local temperatures (up to 5000 K). The relevant effects of this are: (a) reduction in particle/ molecule size, (b) generation of free radicals facilitating mechanisms/ pathways which involve free radicals and (c) increasing the fluid velocity, which facilitates mass transfer during the reactions.

While ultrasonochemistry is now a well-established area, ultrasonoenzymology has been carried out less extensively. In view of harsh local environments during ultrasonication, it is expected that enzymes in nonaqueous media would be able to withstand ultrasonication better (in view of their higher stability in such media). T hus, it is not surprising that attempts have been made to use ultrasonicators in nonaqueous enzymology. Two distinct types of approaches have been used. Either ultrasonication has been used as a pretreatment step (for the biocatalyst) or the reaction has been carried out in the presence of ultrasonic waves. Both ultrasonic probes and ultrasonic baths have been used in ultrasonoenzymology.

Unfortunately, in many cases, temperature has not been controlled. In such cases, one does not know whether the effects on reaction rates are due to ultrasonication alone or due to the rise in temperature as well. It is also dif. cult for others to reproduce results obtained with such ill-de. ned conditions. In other cases, temperature has been controlled either by use of a water-cooled reactor or interrupting ultrasonication so that the temperature does not rise. In the latter cases, cycles of ultrasonication and cooling have been used.

Vulfson et al. carried out subtilisin-catalyzed interesteri . cation and reported (a) that pretreatment of subtilisin suspensions by ultrasound in alcohols led to an increase in enzyme activity. This effect was more pronounced with long-chain alcohols, being 6-8 times more in octanol as compared to the effect observed in short chain alcohols. The effect was dependent both on sonication power and water content of the medium, (b) the enhancement of reaction rates was, however, much higher if the reaction was carried out in the presence of ultrasound.

In hydrocarbons, the more hydrophobic the solvent (greater log P value), the larger was the sonication effect. In the case of organic halides, the relative increases under ultrasonication were $CCl_4 > CH_2Cl_2 > CHCl_3$. The use of ultrasonication enhanced the lipase-catalyzed acylation of a naphthol derivative with vinyl acetate in benzene-ether mixture by 83-fold. No temperature control was applied and hence the enzyme activity decreased if the temperature crossed 37°C.

Bracey et al. have re-examined the issue of the importance of enzyme particle size in nonaqueous environments. Zakfis & Klibanov had categorically stated that 'Internal diffusional limitations were dismissed because ultrasonication of a suspension of chymotrypsin in octane (resulting in a reduction of an average enzyme particle from 270 to 5 lm, as revealed by direct microscopic examination) had no appreciable effect on the enzymatic transesterification rate'. According to Bracey et al., no enhancement in the reaction rate of subtilisin-catalyzed interesterification in hexanol by ultrasonication could be observed.

The attempts by Bracey et al. to reconcile the two contrary results involve discussions on the possible mechanism(s) of ultrasonication effects and thus, are worth describing. Transmission electron microscopy by Bracey et al. found that ultrasonication changed the open honeycomb structure of freeze-dried subtilisin into a plate-like structure. Thus, size reduction did not result in signi. cant increase in surface area. However, the hydration state of the enzyme may be critical.

According to Rozicwski & Russell, solvation of a subtilisin particle in a hydrophobic solvent swells it, increasing the diffusion path for the substrate. If these pores have water molecules, reactant diffusion is inhibited. In such cases, increased . uid velocity by sonication would overcome this inhibition. Bracey et al. believe that their enzyme was drier than those of other workers and hence they failed to observe any increase in reaction rate by ultrasonication. It is also worth noting that it is the hydration level before transferring the enzyme to the organic media, which was found to be critical.

Synthesis and Modification of Polymers

Dordick et al. have used about 80-85% organic solvents as media for peroxidase-catalyzed polymerization of phenols. The use of predominantly nonaqueous media has two advantages: the substrate phenols can be solubilized easily, and unlike in water where dimers and trimers precipitate out of solutions thereby terminating chain growth, oligomers of 1500-2000 Da range could be obtained in quite a few cases. Inter estingly, it was reported that 'the decline in the molecular weight of the polymers above 85% dioxane is presently unclear'. Margolin et al. also reported the lipase-catalyzed polycondensation of reactions between racemic diesters and achiral diols to form optically active trimers and pentamers.

The peroxidase-catalyzed reaction has been further exploited for the synthesis of polyaromatic compounds. Electron paramagnetic resonance studies with spinlabeled heme in peroxidase have revealed that inactivation of horseradish peroxidase (HRP) in anhydrous media is due to the substrate partitioning away from the active site to the bulk solvent. Whereas aromatic polymers were formed in both monophasic and biphasic solvents systems, use of reverse micelles yielded microspheric particles of uniform size (micrometer range or lower).

The size of the microspheres was dependent upon the water to surfactant ratio. These nanoparticles could be used for encapsulating a variety of chemicals and enzymes. Polyaniline s have also been synthesized and derivatized using HRP in solvent-buffer systems. It has been pointed out that these polymers have potential applications in biosensors, optical and electroptical devices and chromatography.

Cellulase has been used in aqueous-cosolvent systems to synthesize crystalline cellulose. Lipases and proteases have been used for the synthesis of polyesters in supercritical fluids. Immobilized lipase has been used to catalyze ringopening/ polymerization of 12-dodecanolide and e-caprolactone. Enzyme-catalyzed poly transesterification of alkanedioates and butane-1,4-diol in organic solvents has also been successfully carried out.

Use of subtilisin (solubilized by formation of proteinsurfactant ion-pair method) in iso-octane for transesteri. cation of a thin layer of amylose (deposited onto zinc-selenide plates) has been described. The modification was followed by FTIR and thermogravimetric analysis of amylose. 1H-NMR showed that all available primary amino groups were acylated. Further work in this area will de. nitely be useful in obtaining novel carbohydrate materials for a variety of applications.

Enzyme Modification by Directed Evolution

This approach originated in the realization that an enzyme is designed or evolved for a particular function and works in vivo as a part of a complex metabolic network. In many situations, these two features are constraints for industrial biocatalysts, especially if they are to be used in unnatural environments such as organic media. The approach consists of:

Random Mutations

At the outset, it is clear that one has to limit the number of mutations to only one or two residues at a time, otherwise the number of sequences generated will be unmanageably large. In the next step, it is possible to recombine and accumulate such beneficial mutations. The choice of a suitable screening procedure also requires careful consideration. The screening procedure has to be simple so that a high throughput screening device such as an ELISA reader can be used. In reality, one desires a combination of new or retained features in the evolved enzyme: high expression levels, suitable pH or temperature optima, enantioselectivity and of course, high activity towards a particular substrate.

While directing evolution one has to make sure that one does not evolve one desirable property at the expense of other(s). A screeningmethod can normally evaluate only one or at the most, a few, features. The overall strategy is then to carry out many generations of random mutagenesis by introducing errors while performing PCR. At each generation, the candidate having the best feature (which is being screened) is subjected to the next generation. The final enzyme, the result of this evolutionary progress, will be a more suitable biocatalyst for predecided function and milieu.

In Vitrocombination

In this alternate way of accumulating desirable changes (and features), the useful mutations identi. ed during a generation can be recombined by DNA shuffling or sexual PCR. When combined with screening, DNA shuf. ing allows the discovery of the best combinations of mutations and does not assume that the best combination contains all the mutations in a population. This form of recombination can simultaneously remove mutations associated with wrong features which may not have been screened out.

There are two approaches to obtain an esterase for hydrolyzing *p*-nitrobenzyl ester in the presence of polar organic solvents. The application was to selectively remove *p*-nitrobenzyl groups during the synthesis of cephalosporin type antibiotics. The starting enzyme was from Bacillus subtilis with low *p*-nitrobenzyl esterase activity, especially in the polar organic solvents required to solubilize the esters. The clone obtained by directed evolution showed more than 150-fold greater total activity than the wild type clone. It was also more stable in 15% (v/v) dimethyl formamide. Unlike site-directed mutagenesis, directed evolution can be applied without much structural and mechanistic information about an enzyme.

Using Antibodies in Organic Solvents

Using antibodies in organic media provides a good opportunity to carry out immunoassays of water insoluble antigens. One major application of this is in environmental control and analysis. Russ ell et al. showed that antigen-antibody interaction is possible in low water containing organic solvents. Stocklein et al. studied binding of triazine herbicides to antibodies in nearly anhydrous media. The work of Matsuura et al. on screening monoclonals for okadaic acid in the context of shell. sh poisoning is worth mentioning. Okahata & Yamaguchi described a lipid-coated catalytic antibody which was soluble in organic cosolvents and had higher kcat as compared to the nativemolecule in suchmedia. Recently, catalytic antibody, 38C2 with aldolase activity, has become commercially available. It has been found that 38C2 can catalyze a retroaldolization reaction in both aqueous and organic solvents.

Hydrolysis and Esterification of Organometallic Substrates

Suitable substrates for this type of reaction include orgnosilicon and organotin compounds and P-bonded complexes of iron, chromium and manganese. A critical feature is allowing stereoselective reactions of water sensitive compounds to be carried out. Kinetic resolution of α-hydroxystannanes by lipases is described. α-Hydroxystannanes are valuable synthetic reagents, notable as precursors to α-alkoxy lithium species, which react with a number of electrophiles, with retention of configuration at the α-center. Several applications include simple quenching with CO_2 to access α-hydroxy acid and double bond transmetallation to generate a functionalized organocopper reagent for use in conjugate addition reactions.

6

Carbohydrate-Processing Phosphorolytic Enzymes

Enzymes involved in the formation and cleavage of glycosyl linkages are categorized into three main classes; hydrolytic enzymes, phosphorolytic enzymes, and synthetic enzymes (glycosyl-nucleotide glycosyltransferases) (Fig. 1). Hydrolytic enzymes are the most extensively studied of the three classes of enzymes, not only in terms of their basic properties but also in terms of their applications. These enzymes are frequently employed in the hydrolysis of polysaccharides and they are very important industrially in the utilization of starch. The reactions of hydrolytic enzymes are substantially irreversible due to the existence of an abundance of water molecules in the reaction mixture.

The synthase enzymes (glycosyl-nucleotide glycosyltransferases) are biologically important because they perform the role of synthesizing carbohydrate chains in vivo. They have also been studied in some detail, mainly with regard to their importance from a biological perspective. The reactions of the synthases are irreversible in the synthetic direction due to the requirement for cleavage of the high-energy bond of the glycosyl-nucleotide (a phosphate diester linkage) in the reaction.

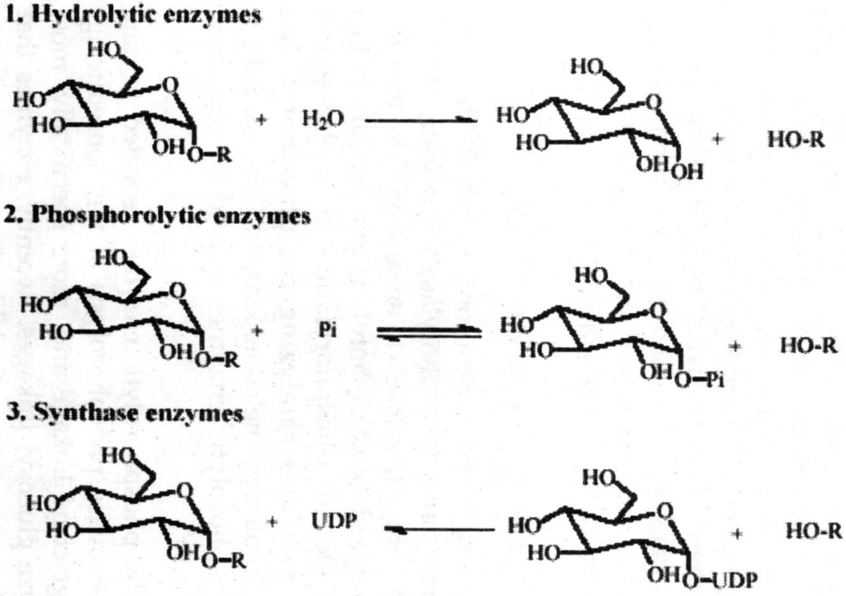

Figure 1. Reactions catalyzed by the three categories of enzymes that involve glycosyl linkages. The reactions illustrated are of 1, α–glucosidase; 2, phosphorylase; 3, starch synthase.

Additionally, most of the synthases are membrane-associated proteins and are too unstable to use in a practical sense in the in vitro synthesis of carbohydrate chains.

The phosphorolytic enzymes however have properties that lie somewhere in between those of the hydrolytic enzymes and the synthases. Since the bond energy of the substrate, a glycosyl-phosphate, is not as high as that of a glycosyl-nucleotide, their reactions are reversible. Therefore, phosphorolytic enzymes can be employed in the practical synthesis of sugar chains, exploiting their reverse reactions. To date however, there have not been as many publications the phosphorolytic enzymes as there have been for the other two classes of enzymes.

Phosphorolytic Enzymes Classification

The enzymes are named using a combination of "the name of the substrate" and "phosphorylase". However, the use of the word "phosphorylase" is an anomaly in that the reactions catalyzed by phosphorolytic enzymes are in fact phosphorolysis, not phosphorylation. It is also worth noting that enzymes catalyzing the phosphorylation reaction are denoted by the name kinases. Thus, we believe that the phosphorolytic enzymes should be re-named "phosphorolases".

All of the phosphorolytic enzymes reported to date catalyze an exowise phosphorolysis at the nonreducing-end of the glycosyl linkage and, more specifically, most phosphorolyze glucosyl linkages. Recently, enzymes that phosphorolyze galactosyl linkages (1) and N-acetylglucosaminyl linkages (2) have been reported. The regiospecificities of the phosphorolytic enzymes are very strict and they phosphorolyze only their specific type of glycosyl linkage. However, these strict specificities are important in the exploitation of these enzymes in the synthesis of oligosaccharides via their reverse reactions. In other words, oligosaccharides with specific glycosyl linkages can be synthe-sized using phosphorolytic enzymes.

Phosphorolytic enzymes are classified by the anomeric forms of the glycoside phosphorolyzed or by the anomeric forms of the glycosyl-1-phosphate produced. Another way used to classify these enzymes is by describing them in terms of the anomeric retention or inversion process they perform in the reaction. These classifications are also those that are applied to the hydrolytic enzymes.

Phosphorylase is the enzyme that catalyzes the reversible phosphorolysis of α–1,4-glucans at the non-reducing end, for example glycogen and amylopectin, to release an α–glucose-1-phosphate. The phosphorolytic enzymes are summarized in Table 1. This enzyme is the most extensively studied of the phosphorolytic enzymes and it is found in animals, plants, and microorganisms.

The role of the enzyme is considered to be in the utilization of storage polysaccharides (glycogen or amylopectin) in the glycolytic pathway. Phosphorylase has a strict requirement for a pyridoxal phosphate and the apo-enzyme does not display any activity. From all sources, these enzymes exist as dimers of an approximately 100 kDa sized protein.

The reaction mechanism of phosphorylase is a rapid equilibrium random bi bi mechanism and in animal muscle, these enzymes are regulated by a phosphorylation/ desphosphorylation system comprising a phosphorylase kinase and a phosphorylase phosphatase. The phosphorylated and dephosphorylated forms are denoted as phosphorylase a and phosphorylase b respectively, and phosphorylase b has a requirement for AMP to display activity. In contrast, phosphorylases

Table I. The phosphorolytic enzymes

EC	name	mechanism	substrate	product	d.p. substrate	family
2.4.1.1	(glycogen) phosphorylase	retention	Glc-α1,4-Glc	α-Glc-1P	>=5 or 4	GT35
2.4.1.7	sucrose phosph- orylase	retention	Glc-α1,_2-Fru	α-Glc-1P	2 only	GH13
2.4.1.8	maltose phosphor- ylase	inversion	Glc-α1,4-Glc	β-Glc-1P	2 only	GH65
2.4.1.20	cellobiose phosph- orylase	inversion	Glc-β1,4-Glc	α-Glc-1P	2 only	GT36
2.4.1.30	laminaridextrin phosphorylase	inversion	Glc-β1,3-Glc	α-Glc-1P	>=2	not cloned
2.4.1.31	laminaribiose phosphorylase	inversion	Glc-β1,3-Glc	α-Glc-1P	>=2	not cloned
2.4.1.49	cellodextrin phosphorylase	inversion	Glc-β1,4-Glc	α-Glc-1P	>=3	GT36
2.4.1.64	trehalose phosphorylase	inversion	Glc-α1,α1-Glc	β-Glc-1P	2 only	not identified
2.4.1.97	β1,3glucan phosphorylase	inversion	Glc-β1,3-Glc	α-Glc-1P	polymer	not cloned
2.4.1.?	trehalose					

Contd . . .

Contd . . .

	phosphorylase	retention	Glc-α1,α1-Glc	α-Glc-1P	2 only	not identified
2.4.1.?	kojibiose phosphorylase	inversion	Glc-α1,2-Glc	β-Glc-1P	>=2	not identified
2.4.1.?	β 1,3Gal-GlyNAc phosphorylase	inversion	Gal-β1,3-GlyNAc	α-Gal-1P	2 only	not cloned
2.4.1.?	chitobiose phosphorylase	inversion	GlcNAc-β1,4-GlcNAc	α-GlcNAc-1P	2 only	GT36

isolated from plants and microorganisms are not regulated by this system. The smallest substrate typically accepted for phosphorolysis by a phosphorylase enzyme is a maltopentaose sugar. After the complete phosphorolysis of amy-lopectin, this enzyme leaves behind a "phosphorylase-limited" dextrin with four α–1,4 linked glucose residues at the branching points. Recently however, it was reported that the smallest substrate accepted by a phosphorylase isolated from a thermophilic bacteria source was maltotetraose, in contrast with that observed for other enzymes.

All of the phosphorylases reported to date are classified into the glycosyltransferase family 35 based on their amino acid sequences. The 3D structure of the phosphorylase isolated from rabbit muscle was solved in 1979 (9,10), and the 3D structures of the E. coli, yeast, and human (PDB/3D accession numbers; 1EM6, 1EXV, 1FA9, 1FC0) enzymes are now available.

Sucrose phosphorylase catalyzes the reversible phosphorolysis of sucrose into α–glucose-1-phosphate and fructose. This enzyme is found in bacterial cells and is thought to be involved in the metabolism of extracellular sucrose. The phosphorolysis reaction proceeds via a ping pong bi bi mechanism and the enzyme transfers an α–glucoside in the absence of phosphate.

Using this enzyme in a transglucosyl manner, several α–glucosides were able to be synthesized in one step from sucrose and various acceptor molecules. Taking into account the reaction mechanism, sucrose phosphorylase can be considered to be an α–glucosyl transferase that also transfers glucose to phosphate- quite different from the other phosphorolytic enzymes. This apparent anomaly is supported by the fact that sucrose phosphorylases are classified into glucosyl hydrolase family 13 based on their amino acid sequences. Family 13 contains hydrolytic enzymes such as α–amylase, α–glucosidase and glycosyl-transferring enzymes such as cyclodextrin glucanotransferase and α–1,4-glucanotransferase. Sucrose phosphorylase is used in a diagnostic reagent for the quantification of phosphate in blood and the colorimetric quantification of sucrose using this enzyme has also been reported.

Maltose phosphorylase reversibly phosphorolyzes maltose into β–glucose-1-phosphate and glucose.This enzyme has been found in bacterial cells and is thought to be involved in the metabolism of extracellular maltose. The β–glucose-1-phosphate generated by the enzyme is converted into glucose-6-phosphate by the action of a β–specific phosphoglucomutase to facilitate entry into the glycolytic pathway. The reaction mechanism for maltose phosphorylase is reported to be a sequential bi bi mechanism. Since maltose phosphorylase recognizes the α–anomeric hydroxyl group at the reducing end of maltose, the enzyme phosphorolyzes only the disaccharide unit. Exploiting this fact, a specific determination of maltose employing maltose phosphorylase has been reported.

Maltose phosphorylases are classified into glycosyl hydrolase family 65 along with trehalases. Recently, the 3D structure of the maltose phosphorylase isolated from Lactobα–cillus brevis was described. Cellobiose phosphorylase catalyzes the reversible phosphorolysis of cellobiose into α–glucose-1-phosphate and glucose. These enzymes are present as intracellular enzymes in bacterial cells that utilize cellulose. Cellobiose phosphorylase is thought to be involved in the utilization of extracellular cellobiose formed by the action of cellulase.

The reaction mechanism for this enzyme is reported to be an ordered bi bi process. This enzyme recognizes the β–anomeric hydroxyl group at the reducing end of the cellobiose moiety, and this has been confirmed through experiments employing pseudo-glucoses. Therefore, this enzyme does not phosphorolyze cellooligosaccharides larger than cellotriose. The acceptor specificity in the reverse reaction has also been investigated and it was found that the enzyme has a strict recognition of the hydroxyl groups of the glucose molecule at positions β–1, 3, and 4, but not of the hydroxyls at positions 2 and 6. As such, several heterooligosaccharides have been synthesized based on these specificities.

Cellobiose phosphorylases are classified into glycosyltransferase family 36 along with cyclic-β–1,2-glucan synthase, cellodextrin phosphorylase and chitobiose phosphorylase. β–1, 3- Oligoglucan phosphorylase, laminaribiose phosphorylase, and β–1,3-glucan phosphorylase phosphorolyze β–1,3-glucosidic linkages to generate α–glucose-1-phosphate. These enzymes have been found as intracellular enzymes in algae and euglenida cells and the role of these enzymes is thought to be in the utilization of the storage polysaccharide paramylon (β–1,3-glucan). β–1,3-Oligoglucan phosphorylase and laminaribiose phosphorylase were found in Euglena gracilis cells and both of these enzymes phosphorolyze a series of laminarioligosaccharides.

The key difference in the reactions catalyzed by the two enzymes is that β–1, 3-oligoglucan phosphorylase phosphorolyzes laminaritriose faster than laminaribiose and vice versa for laminaribiose phosphorylase. Another difference is that β–1,3-oligoglucan phosphorylase requires a thiol compound whereas laminaribiose phosphorylase does not. β–1, 3-Glucan phosphorylase was purified from a cell-free extract of Ochromonas malhamensis. In contrast to the other two enzymes, this enzyme phosphorolyzes a polymeric substrate.

The reaction mechanism catalyzed by laminaribiose phosphorylase was determined to occur in an ordered bi bi process and in the reverse reaction, the enzyme recognized various monosaccharides and β–glucosides as acceptor molecules. Mixtures of laminarioligosaccharides having varying degrees of polymerization have been synthesized from glucose and α–glucose-1-phosphate using these enzymes.

Cellodextrin phosphorylase catalyzes the reversible phosphorolysis of cellooligosaccharides larger than cellotriose, to generate α–glucose-1-phosphate. This enzyme has only been found in the cells of Clostridia, accompanied with cellobiose phosphorylase. The regio-specificity of cellodextrin phosphorylase is identical with that of cellobiose phosphorylase but the specificity with regards to the degree of polymerization of the acceptor is quite different. In the reverse reaction, the enzyme does not use glucose as the acceptor but recognizes various aryl-β–glucosides and β–glucosyl disaccharides.

Analogues of cellooligosaccharides that act as cellulase inhibitors have been synthesized using cellodextrin phosphorylase. These enzymes are classified into glycosyltransferase family 36 along with cellobiose phosphorylase. Trehalose phosphorylase reversibly phosphorolyzes trehalose to form β–glucose-1-phosphate and glucose. This enzyme was first discovered in the cells of Euglena gracilis and has been implicated in the metabolism of intracellular trehalose, a storage disaccharide present in the strain. In the 1990s, trehalose phosphorylases were discovered in various bacteria. The synthesis of trehalose from maltose using maltose phosphorylase in conjunction with this enzyme has been investigated.

Trehalose phosphorylase (retention type) reversibly catalyzes the phosphorolyis of trehalose into α–glucose-1-

phosphate and glucose. The retention type enzyme was isolated from various fungi cells, edible mushrooms included. The reaction mechanism of this enzyme was described as being a sequential bi bi process, different from that of sucrose phosphorylase.

Trehalose phosphorylase does not catalyze transglucosyl reactions in the absence of phosphate. Kojibiose phosphorylase was discovered in the cell-free extract of a thermophilic bacterium accompanied by trehalose phosphorylase (inversion type). This enzyme catalyzes the reversible phosphorolysis of kojibiose into β–glucose-1-phosphate and glucose. In the reverse reaction, the enzyme utilizes various monosaccharides and oligosaccharides as acceptors.

β–1, 3-Gal-GlyNAc phosphorylase was isolated as an intracellular enzyme from Bifidobacterium bifidus. This enzyme phosphorolyzes β–1, 3-Gal-GlyNAc into α–galactose-1-phosphate and GlyNAc. In the reverse reaction, the enzyme specifically recognizes α–galactose-1-phosphate as the donor and only N-acetylglucosamine and N-acetylgalactosamine as acceptors. Several derivatives of β–1, 3-Gal-GlyNAc have been synthesized employing this enzyme.

Chitobiose phosphorylase phosphorolyzes the conversion of chitobiose into α–N-acetylglucosamine-1-phosphate and Nacetylglucosamine. This enzyme was discovered in the cells of a marine bacterium, Vibrio furnissii. The amino acid sequence of β–1,3-Gal-GlyNAc phosphorylase has 87% identity with that of a gene of V. chorelae, which is described as a cellobiose phosphorylase. The role of β–1, 3-Gal-GlyNAc phosphorylase is thought to be in the utilization of extracellular chitobiose formed by the action of chitinase. This enzyme is classified into glycosyltransferase family 36 along with cellobiose phosphorylase and cellodextrin phosphorylase.

Reactions of Cellobiose Phosphorylase

The phosphorolytic reaction was found to proceed via an ordered bi mechanism. The order of the substrate binding and the release of the products were found to be as follows: initially cellobiose bound the enzyme followed by phosphate, then glucose was released and, finally, glucose-1-phosphate was released to complete the reaction (Fig. 2). In addition to glucose, the enzyme recognized a number of monosaccharides as acceptors to form various β–1,4-glucosyl heterodisaccharides. The acceptor specificity of cellobiose phosphorylase is given in Table 2. It was found that the enzyme strictly recognizes the hydroxyl groups at positions 3 and 4 but it is not as discriminating at positions 2 and 6 (Fig. 3).

Figure 2. Ordered bi bi mechanism in the phosphorolysis catalyzed by cellobiose phosphorylase.

In the reverse process, the reaction using β–glucose as the acceptor was significantly faster than that obtained using α–glucose. Furthermore, the enzyme did not accept 1,5-anhydroglucitol (1-deoxyglucose) or β–methyl glucoside, suggesting that the enzyme recognized the β–anomeric hydroxyl group of the acceptor molecule. To confirm this recognition pattern, two pseudo-glucoses (Fig. 4), possessing the structure of a glucopyranose but with the ring oxygen substituted for a methylene group, were examined as acceptors.

The enzyme recognized only the pseudo-β–glucopyranose and not the pseudo-α–glucopyranose, indicating that the enzyme strictly recognizes the β–anomeric hydroxyl group of the acceptor molecule (Fig.

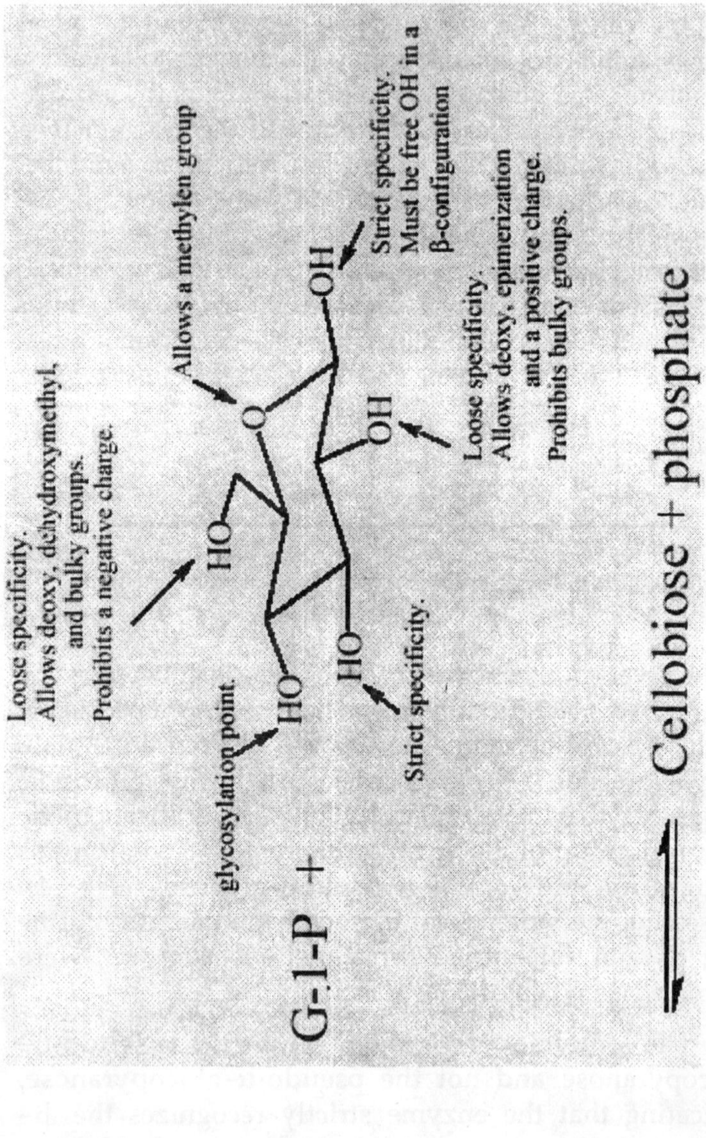

Figure 3. Summary of the acceptor specificity of cellobiose phosphorylase

3). It was postulated that disaccharides having a structure comprising of a "glucose bound by another saccharide at position 6" could serve as the acceptor due to the loose specificity at position 6. Thus, the disaccharides isomaltose, gentiobiose and melibiose were examined as acceptors.

α-glucopyranose β-glucopyranose

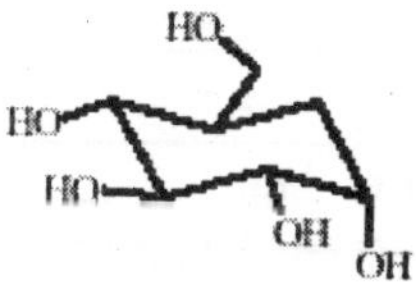

5a-carba-α-glucopyranose (pseudo-α-glucopyranose)

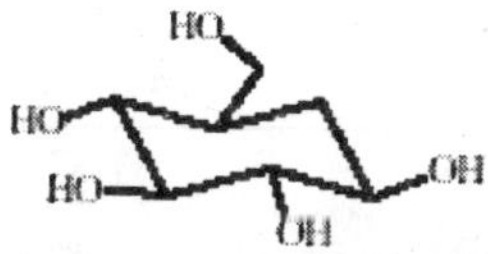

5a-carba-β-glucopyranose (pseudo-β-glucopyranose)

Figure 4. Structures of the pseudo-glucoses.

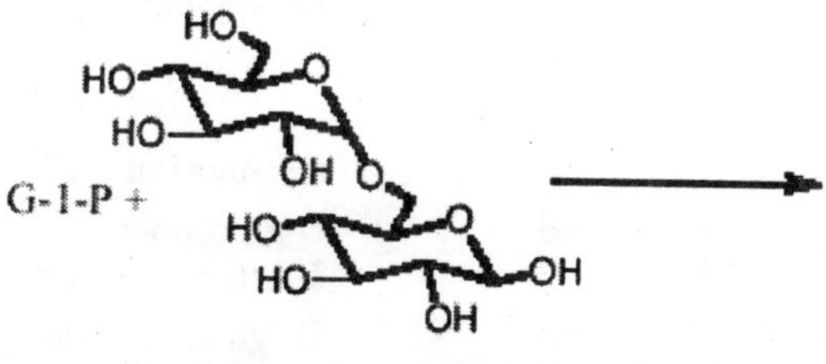

isomaltose (Glcα1-6Glc)

Contd . . .

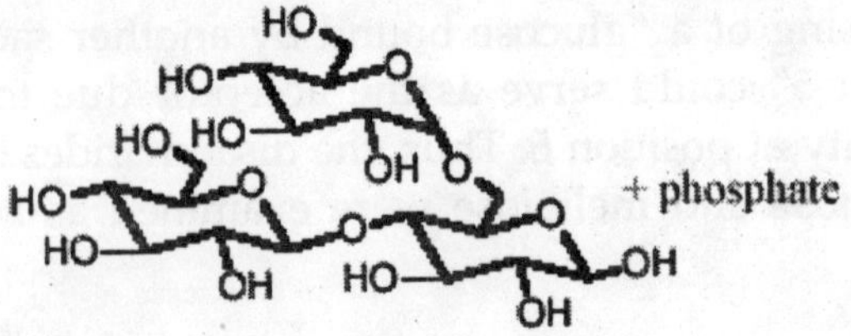

Glcβ1-4(Glcα1-6)Glc

Figure 5. Synthesis of branched trisaccharides using cellobiose phosphorylase.

Table 2. Acceptor specificity of cellobiose phosphorylase

substrate	v (μmol/min/U)	substrate	v (μmol/min/U)
D-glucose	1.37	D -fructose	nd
α-methyl- D -glucoside	nd	L-sorbose	nd
β-methyl- D -glucoside	nd	D -lyxose	nd
1,5-anhydro- D -glucitol	nd	D -ribose	nd
D -mannose	0.054	L -arabinose	nd
2-deoxy- D -glucose	0.034	D -arabinose	nd
D -glucosamine	0.12	L -xylose	nd
N-acetyl- D-glucosamine	nd	*myo*-inositol	nd
D -allose	nd	D-glucono- δ-lactone	nd
D -galactose	nd	D -glucitol	nd
L -idose	nd	D-mannitol	nd
6-deoxy- D-glucose	0.70		
D -glucuronic acid	nd		
D -xylose	0.055		

Determined using 10 mM G-1-P and 10 mM acceptor. nd; <0.03.

Cellobiose phosphorylase transferred a β–glucosyl residue at the reducing-end to position 4 of the glucose moiety of the respective disaccharides to form the corresponding branched trisaccharides as expected (Fig. 5). In the reverse reaction, substrate inhibition was observed using glucose, glucosamine and 6-deoxyglucose as the acceptors, with glucose displaying the strongest level of inhibition.

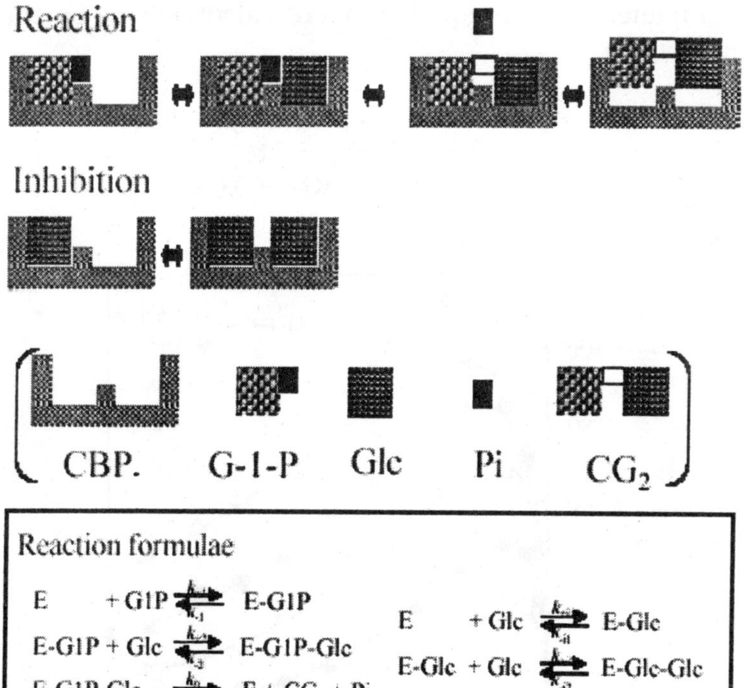

Reaction formulae

$$E + G1P \underset{k_{-1}}{\overset{k_{+1}}{\rightleftharpoons}} E\text{-}G1P$$
$$E\text{-}G1P + Glc \underset{k_{-2}}{\overset{k_{+2}}{\rightleftharpoons}} E\text{-}G1P\text{-}Glc$$
$$E\text{-}G1P\text{-}Glc \xrightarrow{k_0} E + CG_2 + Pi$$
$$E + Glc \underset{k_{-i1}}{\overset{k_{+i1}}{\rightleftharpoons}} E\text{-}Glc$$
$$E\text{-}Glc + Glc \underset{k_{-i2}}{\overset{k_{+i2}}{\rightleftharpoons}} E\text{-}Glc\text{-}Glc$$

Rate equation

$$v = \frac{k_0[E]_0[A][B]}{K_{iA}K_{mB} + \left(K_{mA} + \frac{K_{iA}K_{mB}}{K_{i1}}\right)[B] + K_{mB}[A] + [A][B] + \left(\frac{K_{iA}K_{mB}}{K_{i1}K_{i2}} + \frac{K_{mA}}{K_{i1}}\right)[B]^2 + \frac{K_{mA}}{K_{i1}K_{i2}}[B]^3}$$

Figure 6. Mechanism of competitive substrate inhibition

It was assumed that the phenomenon was caused by the competitive inhibition of the substrate glucose toward glucose-1-phosphate, the other substrate, namely a "competitive substrate inhibition" (Fig. 6). The rate equation constructed using this inhibition model is also given in Fig. 6. The experimental data were regressed with this equation using the Gauss-Newton method to minimize the summation of the square of errors, and the kinetic parameters in the equation were calculated to be the following:

k_{cat} = 120 sec^{-1}, K_{iA} = 0.5 mM , K_{mA} = 2.0 mM, K_{mB} = 2.8 mM, K_{I1} = 170 mM, KI2 = 3.0 mM.

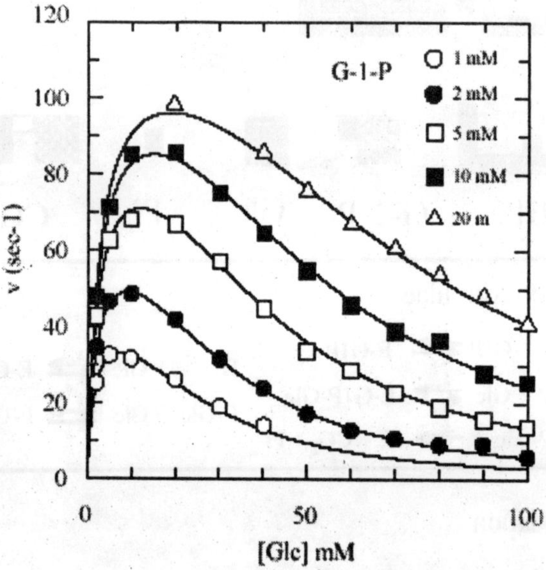

Figure 7. [Glc]-v curve of the reverse reaction catalyzed by ***cellobiose phosphorylase.***

The theoretical curves fitted well with the experimental data (Fig. 7), suggesting that the substrate inhibition model given in Fig. 6 describes the observed phenomenon well. In the reverse reaction, a similar inhibition by glucose was observed and was explained using the same "competitive substrate inhibition" model. It was postulated that the phenomenon would be commonly observed in the reverse reaction of phosphorolytic enzymes which phosphorolyze a glucobiose moiety.

Using Synthesis of Oligosaccharides

Synthesis of 4-O-β-D-Glucopyranosyl-D-Xylose

This microorganism was grown in a medium with CMC as a carbon source and the cells were collected by centrifugation followed by a cold acetone treatment to obtain acetone-dried cells containing cellobiose phosphorylase. However, the dried cells also contained phosphoglucomutase and the contamination caused a considerable decrease in the yield of the cellobiose phosphorylase reaction due to its competing reaction of converting glucose-1-phosphate into glucose-6-phosphate. Therefore, the phosphoglucomutase was selectively removed through treatment of the acetone-dried cells with 50% ethanol solution (Table 3).

Table 3. Activity of cellobiose phosphorylase and phosphoglucomutase in acetone-dried cells- before and after treatment with 50% ethanol

treatment	cellobiose phosphorylase (IU/g cells)	phosphoglucomutase (IU/g cells)
before	2.1	6.1
after	1.3	0.1

Using the acetone-ethanol treated cells as the catalyst, a heterodisaccharide, GX, was synthesized in 60% yield from equimolar quantities of glucose-1-phosphate and xylose. This yield was identical with that obtained using the purified enzyme. GX was isolated by charcoal chromatography and its structure was elucidated. From the data obtained from the enzymatic hydrolysis, ie methylation analysis, FAB-MS, 1H and 13C NMR, the structure of GX was determined to be 4-O-β–D-glucopyranosyl-D-xylose. The structure of GX is intermediate to those of cellobiose and xylobiose, and the characteristics of GX were also intermediate. The β–pnitrophenyl glycoside of GX was chemically synthesized and found to be a useful substrate in the characterization of xylanases.

Synthesis of Laminarioligosaccharides

Mixtures of laminarioligosaccharides with varying degrees of polymerization were synthesized from glucose-1-phosphate and glucose by using a cell-free extract of Euglena gracilis containing laminaribiose phosphorylase and β–1,3-oligoglucan phosphorylase. The average degree of polymerization increased in concert with an increase in the glucose-1-phosphate / glucose ratio. With a ratio of 1 and 20, laminarioligosaccharides with a d.p. of 1-9 and 2-14 were detected by HPLC and the average degrees of polymerization were 1.8 and 6.3, respectively. Using this method, laminarioligosaccharides larger than laminaridecaose were able to be synthesized, which are difficult to obtain using the standard method which involves the limited hydrolysis of β–1,3-glucan.

Conversion of Sucrose into Cellobiose

Cellobiose can be synthesized from glucose-1-phosphate and glucose using cellobiose phosphorylase (Fig. 8, Scheme 1). However, this reaction has its limitations as a practical process because glucose-1-phosphate is prohibitively expensive. However, sucrose phosphorylase can be used to generate glucose-1-phosphate because this enzyme can phosphorolyze a cheap substrate, sucrose, into fructose and glucose-1-phosphate, the anomeric configuration of which is identical (a) with that used by cellobiose phosphorylase.

Thus, a reaction system to form cellobiose and fructose using sucrose and glucose as the raw materials along with a catalytic amount of phosphate can be constructed using the combined action of sucrose phosphorylase and cellobiose phosphorylase (Fig. 8, Scheme 2). However, this reaction system still has two problems: 1. the formation of an equimolar amount of fructose as a byproduct; and 2. the competitive inhibition by glucose toward glucose-1-phosphate for both enzymes (competitive substrate inhibition for cellobiose phosphorylase) causing a considerable decrease in the activities of the two enzymes. To solve these problems, the addition of a third enzyme to the system, xylose isomerase, was evaluated.

Scheme 1

G-1-P + Glc ⇄ Cellobiose + Pi (CBP)

Scheme 2

Suc + Pi ⇄ G-1-P + Fru (SP)

G-1-P + Glc ⇄ Cellobiose + Pi (CBP)

Suc + Glc ⇄ Cellobiose + Fru

(Two enzymes and a catalytic amount of Pi)

Scheme 3

Suc + Pi ⇄ G-1-P + Fru (SP)

Fru ⇄ Glc (XI)

G-1-P + Glc ⇄ Cellobiose + Pi (CBP)

Suc ⇄ Cellobiose

(Three enzymes and a catalytic amount of Pi)

Figure 8. Syntheses of cellobiose via the reverse reaction of cellobiose phosphorylase CBP, cellobiose phosphorylase; SP, sucrose phosphorylase; XI, xylose isomerase.

Xylose isomerase is used industrially to produce high fructose corn syrup by the isomerization of glucose into fructose. In a reaction mixture containing sucrose, a catalytic amount of phosphate, and the three enzymes (sucrose phosphorylase, xylose isomerase and cellobiose phosphorylase), the following reactions were anticipated: the initial phosphorolysis of sucrose into glucose-1-phosphate and fructose by sucrose phosphorylase. The fructose is then converted into glucose by xylose isomerase.

Finally, cellobiose and phosphate is generated from the glucose-1-phosphate and the glucose by cellobiose phosphorylase. Since the phosphate is recycled in the reaction of sucrose phosphorylase it was predicted that sucrose would be converted into cellobiose in one step (Fig. 8, Scheme 3). The system was examined and it was found that cellobiose was able to be synthesized from sucrose in the presence of a catalytic amount of phosphate as predicted. The reaction yield of cellobiose was 76%, higher than the equilibrium yield using glucose-1-phosphate as the starting material (64%).

A. B.

Figure 9. Auto-crystallization of cellobiose in the conversion of sucrose into cellobiose. A, before the reaction; B, after 8 days reaction. The reaction was performed at pH 7.0 at 30°C containing 500 g/L sucrose, 2 IU/mL sucrose phosphorylase, 1 IU/mL xylose isomerase and 4 IU/mL cellobiose phosphorylase.

The higher yield is considered to be attributable to the high energy of β–fructofuranosyl linkage in the sucrose molecule. The cellobiose thus produced was able to be isolated using a charcoal column and was crystallized to give an overall yield of 42%. The three-enzyme system is applicable in the industrial production of cellobiose because of the high yield and the cheap raw material (sucrose). Recently, we found that the generated cellobiose auto-crystallized in the reaction mixture when starting from a high concentration of sucrose such as 500 g/ L (Fig. 9). Since it is possible to separate the cellobiose produced from the reaction mixture by exploiting this phenomenon, a semi-continuous production of cellobiose using this system is possible.

Conversion of Sucrose into Laminaribiose

Sucrose was converted into laminaribiose in one step by the combined action of sucrose phosphorylase, xylose isomerase and laminaribiose phosphorylase in the presence of a catalytic amount of phosphate. Interestingly, laminaritriose was formed as a byproduct in this enzyme system whereas no cellotriose was formed during the corresponding synthesis of cellobiose. This difference was caused by the relative specificities of the enzymes; cellobiose phosphorylase does not use cellobiose as an acceptor whereas laminaribiose phosphorylase can use laminaribiose as an acceptor to form laminaritriose. The yields of laminaribiose and laminaritriose were 55% and 11%, respectively.

7

Biosensors

Enzymes Analysis

Enzymes make excellent analytical reagents due to their specificity, selectivity and efficiency. They are often used to determine the concentration of their substrates (as analytes) by means of the resultant initial reaction rates. If the reaction conditions and enzyme concentrations are kept constant, these rates of reaction (v) are proportional to the substrate concentrations ([S]) at low substrate concentrations. When $[S] < 0.1\ K_m$, equation

$$v = (V_{max}/K_m)[S] \quad (1)$$

The rates of reaction are commonly determined from the difference in optical absorbance between the reactants and products. An example of this is the β-D-galactose dehydrogenase assay for galactose which involves the oxidation of galactose by the redox coenzyme, nicotine-adenine dinucleotide (NAD^+).

$$\beta\text{-D-galactose} + NAD^+ \longrightarrow \text{D-galactono-1,4-lactone}^+\ NADH^+\ H^+$$

A 0.1 mM solution of NADH has an absorbance at 340nm, in a 1 cm path-length cuvette, of 0.622, whereas the NAD^+ from which it is derived has effectively zero absorbance at this wavelength. The conversion (NAD^+ NADH) is, therefore, accompanied by a large increase in absorption of light at this wavelength. For the reaction to be linear with respect to the galactose concentration, the galactose is kept within a concentration range well below the Km of the enzyme for galactose.

In contrast, the NAD^+ concentration is kept within a concentration range well above the K_m of the enzyme for NAD^+, in order to avoid limiting the reaction rate. Such assays are commonly used in analytical laboratories and are, indeed, excellent where a wide variety of analyses need to be undertaken on a relatively small number of samples.

The drawbacks to this type of analysis become apparent when a large number of repetitive assays need to be performed. Then, they are seen to be costly in terms of expensive enzyme and coenzyme usage, time consuming, labour intensive and in need of skilled and reproducible operation within properly equipped analytical laboratories. For routine or on-site operation, these disadvantages must be overcome. This is being achieved by the production of biosensors which exploit biological systems in association with advances in micro-electronic technology.

A biosensor is an analytical device which converts a biological response into an electrical signal (Figure 1). The term 'biosensor' is often used to cover sensor devices used in order to determine the concentration of substances and other parameters of biological interest even where they do not utilise a biological system directly. This very broad definition is used by some scientific journals but will not be applied to the coverage here.

Enzymes as the biologically responsive material, but it should be recognised that other biological systems may be utilised by biosensors, for example, whole cell metabolism, ligand binding and the antibody-antigen reaction.

Biosensors represent a rapidly expanding field, at the present time, with an estimated 60% annual growth rate; the major impetus coming from the health-care industry (e.g. 6% of the western world are diabetic and would benefit from the availability of a rapid, accurate and simple biosensor for glucose) but with some pressure from other areas, such as food quality appraisal and environmental monitoring.

The estimated world analytical market is about £12,000,000,000 year-1 of which 30% is in the health care area. There is clearly a vast market expansion potential as less than 0.1% of this market is currently using biosensors. Research and development in this field is wide and multidisciplinary, spanning biochemistry, bioreactor science, physical chemistry, electrochemistry, electronics and software engineering.

Most of this current endeavour concerns potentiometric and amperometric biosensors and colorimetric paper enzyme strips. However, all the main transducer types are likely to be thoroughly examined, for use in biosensors, over the next few years. A successful biosensor must possess at least some of the following beneficial features:

1. The biocatalyst must be highly specific for the purpose of the analyses, be stable under normal storage conditions and, except in the case of colorimetric enzyme strips and dipsticks, show good stability over a large number of assays (i.e. much greater than 100).

2. The reaction should be as independent of such physical parameters as stirring, pH and temperature as is manageable. This would allow the analysis of samples with minimal pre-treatment. If the reaction involves cofactors or coenzymes these should, preferably, also be co-immobilised with the enzyme.

3. The response should be accurate, precise, reproducible and linear over the useful analytical range, without dilution or concentration. It should also be free from electrical noise.

4. If the biosensor is to be used for invasive monitoring in clinical situations, the probe must be tiny and biocompatible, having no toxic or antigenic effects. If it is to be used in fermenters it should be sterilisable. This is preferably performed by autoclaving but no biosensor enzymes can presently withstand such drastic wet-heat treatment. In either case, the biosensor should not be prone to fouling or proteolysis.

5. The complete biosensor should be cheap, small, portable and capable of being used by semi-skilled operators.

6. There should be a market for the biosensor. There is clearly little purpose developing a biosensor if other factors (e.g. government subsidies, the continued employment of skilled analysts, or poor customer perception) encourage the use of traditional methods and discourage the decentralisation of laboratory testing.

The biological response of the biosensor is determined by the biocatalytic membrane which accomplishes the conversion of reactant to product. Immobilised enzymes possess a number of advantageous features which makes them particularly applicable for use in such systems. They may be re-used, which ensures that the same catalytic activity is present for a series of analyses.

This is an important factor in securing reproducible results and avoids the pitfalls associated with the replicate pipetting of free enzyme otherwise necessary in analytical protocols. Many enzymes are intrinsically stabilised by the immobilisation process, but even where this does not occur there is usually considerable apparent stabilisation. It is normal to use an excess of the enzyme within the immobilised sensor system. This gives a catalytic redundancy (i.e. $h << 1$) which is sufficient to ensure an increase in the apparent stabilisation of the immobilised enzym.

Even where there is some inactivation of the immobilised enzyme over a period of time, this inactivation is usually steady and predictable. Any activity decay is easily incorporated into an analytical scheme by regularly interpolating standards between the analyses of unknown samples. For these reasons, many such immobilised enzyme systems are re-usable up to 10,000 times over a period of several months. Clearly, this results in a considerable saving in terms of the enzymes' cost relative to the analytical usage of free soluble enzymes.

When the reaction, occurring at the immobilised enzyme membrane of a biosensor, is limited by the rate of external diffusion, the reaction process will possess a number of valuable analytical assets. It follows that the biocatalyst gives a proportional change in reaction rate in response to the reactant (substrate) concentration over a substantial linear range, several times the intrinsic K_{m}.

This is very useful as analyte concentrations are often approximately equal to the Kms of their appropriate enzymes which is roughly 10 times more concentrated than can be normally determined, without dilution, by use of the free enzyme in solution. Control of biosensor response by the external diffusion of the analyte can be encouraged by the use of permeable membranes between the enzyme and the bulk solution. The thickness of these can be varied with associated effects on the proportionality constant between the substrate concentration and the rate of reaction. Even if total dependence on the external diffusional rate is not achieved (or achievable), any increase in the dependence of the reaction rate on external or internal diffusion will cause a reduction in the dependence on the pH, ionic strength, temperature and inhibitor concentrations.

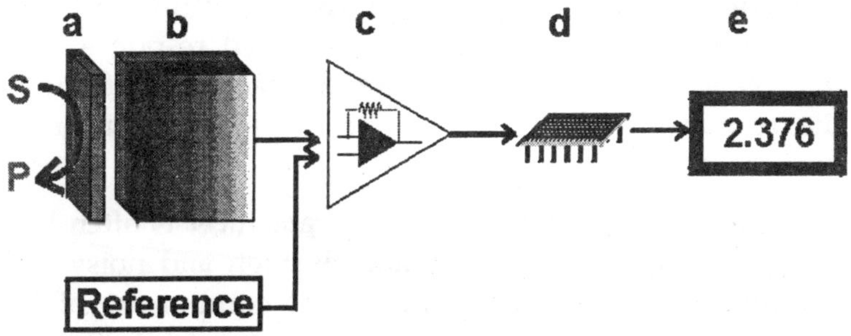

Figure 1. Schematic diagram showing the main components of a biosensor. The biocatalyst (a) converts the substrate to product. This reaction is determined by the transducer (b) which converts it to an electrical signal. The output from the transducer is amplified (c), processed (d) and displayed (e).

The key part of a biosensor is the transducer which makes use of a physical change accompanying the reaction. This may be

1. the heat output (or absorbed) by the reaction (calorimetric biosensors),
2. changes in the distribution of charges causing an electrical potential to be produced (potentiometric biosensors),
3. movement of electrons produced in a redox reaction (amperometric biosensors),

4. light output during the reaction or a light absorbance difference between the reactants and products (optical biosensors), or
5. effects due to the mass of the reactants or products (piezo-electric biosensors).

There are three so-called 'generations' of biosensors; First generation biosensors where the normal product of the reaction diffuses to the transducer and causes the electrical response, second generation biosensors which involve specific 'mediators' between the reaction and the transducer in order to generate improved response, and third generation biosensors where the reaction itself causes the response and no product or mediator diffusion is directly involved.

The electrical signal from the transducer is often low and superimposed upon a relatively high and noisy (i.e. containing a high frequency signal component of an apparently random nature, due to electrical interference or generated within the electronic components of the transducer) baseline.

The signal processing normally involves subtracting a 'reference' baseline signal, derived from a similar transducer without any biocatalytic membrane, from the sample signal, amplifying the resultant signal difference and electronically filtering (smoothing) out the unwanted signal noise.

The relatively slow nature of the biosensor response considerably eases the problem of electrical noise filtration. The analogue signal produced at this stage may be output directly but is usually converted to a digital signal and passed to a microprocessor stage where the data is processed, converted to concentration units and output to a display device or data store.

Calorimetric Biosensors

Many enzyme catalysed reactions are exothermic, generating heat (Table 1) which may be used as a basis for measuring the rate of reaction and, hence, the analyte concentration. This represents the most generally applicable type of biosensor. The temperature changes are usually determined by means of thermistors at the entrance and exit of small packed bed columns containing immobilised enzymes within a constant temperature environment (Figure 2).

Under such closely controlled conditions, up to 80% of the heat generated in the reaction may be registered as a temperature change in the sample stream. This may be simply calculated from the enthalpy change and the amount reacted. If a 1 mM reactant is completely converted to product in a reaction generating 100 kJ mole-1 then each ml of solution generates 0.1 J of heat. At 80% efficiency, this will cause a change in temperature of the solution amounting to approximately 0.02°C. This is about the temperature change commonly encountered and necessitates a temperature resolution of 0.0001°C for the biosensor to be generally useful.

Table 1. Heat output (molar enthalpies) of enzyme catalysed reactions.

Reactant	Enzyme	Heat output $-\Delta H$ (kJ mole^{-1})
Cholesterol	Cholesterol oxidase	53
Esters	Chymotrypsin	4 - 16
Glucose	Glucose oxidase	80
Hydrogen peroxide	Catalase	100
Penicillin G	Penicillinase	67
Peptides	Trypsin	10 - 30
Starch	Amylase	8
Sucrose	Invertase	20

Urea	Urease	61
Uric acid	Uricase	49

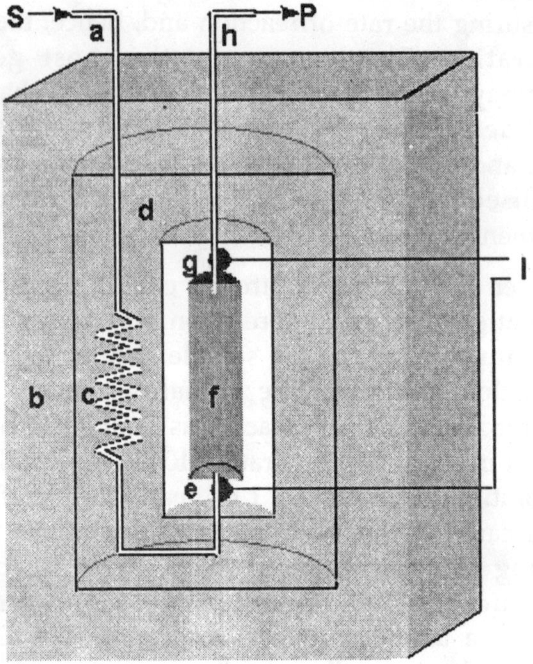

Figure 2. Schematic diagram of a calorimetric biosensor.

The sample stream (a) passes through the outer insulated box (b) to the heat exchanger (c) within an aluminium block (d). From there, it flows past the reference thermistor (e) and into the packed bed bioreactor (f, 1ml volume), containing the biocatalyst, where the reaction occurs. The change in temperature is determined by the thermistor (g) and the solution passed to waste (h). External electronics (l) determines the difference in the resistance, and hence temperature, between the thermistors.

The thermistors, used to detect the temperature change, function by changing their electrical resistance with the temperature, obeying the relationship

$$Ln\left(\frac{R_1}{R_2}\right) = B\left(\frac{1}{T_1} - \frac{1}{T_2}\right) \tag{2}$$

therefore:

$$\frac{R_1}{R_2} = e^{B\left(\frac{1}{T_1} - \frac{1}{T_2}\right)} \tag{2b}$$

where R_1 and R_2 are the resistances of the thermistors at absolute temperatures T_1 and T_2 respectively and B is a characteristic temperature constant for the thermistor. When the temperature change is very small, as in the present case, $B(1/T_1) - (1/T_2)$ is very much smaller than one and this relationship may be substantially simplified using the approximation when $x<<1$ that $e^x \approx 1 + x$ (x here being $B(1/T_1) - (1/T_2)$),

$$R_1 = R_2\left\{1 + B\left(\frac{T_2 - T_1}{T_1 T_2}\right)\right\} \tag{3}$$

As $T_1 \approx T_2$, they both may be replaced in the denominator by T_1.

$$\frac{\Delta R}{R} = -\left(\frac{B}{T_1^2}\right)\Delta T \tag{4}$$

The relative decrease in the electrical resistance (DR/R) of the thermistor is proportional to the increase in temperature (ΔT). A typical proportionality constant ($-B/T_1^2$) is $-4\%°C^{-1}$. The resistance change is converted to a proportional voltage change, using a balanced Wheatstone bridge incorporating precision wire-wound resistors, before amplification. The expectation that there will be a linear correlation between the response and the enzyme activity has been found to be borne out in practice. A major problem with this biosensor is the difficulty encountered in closely matching the characteristic temperature constants of the measurement and reference thermistors.

An equal movement of only 1°C in the background temperature of both thermistors commonly causes an apparent change in the relative resistances of the thermistors equivalent to 0.01°C and equal to the full-scale change due to the reaction. It is clearly of great importance that such environmental temperature changes are avoided, which accounts for inclusion of the well-insulated aluminium block in the biosensor design (Figure 2).

The sensitivity (10^{-4} M) and range (10^{-4} - 10^{-2} M) of thermistor biosensors are both quite low for the majority of applications although greater sensitivity is possible using the more exothermic reactions (e.g. catalase). The low sensitivity of the system can be increased substantially by increasing the heat output by the reaction.

In the simplest case this can be achieved by linking together several reactions in a reaction pathway, all of which contribute to the heat output. Thus the sensitivity of the glucose analysis using glucose oxidase can be more than doubled by the co-immobilisation of catalase within the column reactor in order to disproportionate the hydrogen peroxide produced. An extreme case of this amplification is shown in the next page recycle scheme for the detection of ADP.

ADP is the added analyte and excess glucose, phosphoenol pyruvate, NADH and oxygen are present to ensure maximum reaction. Four enzymes (hexokinase, pyruvate kinase, lactate dehydrogenase and lactate

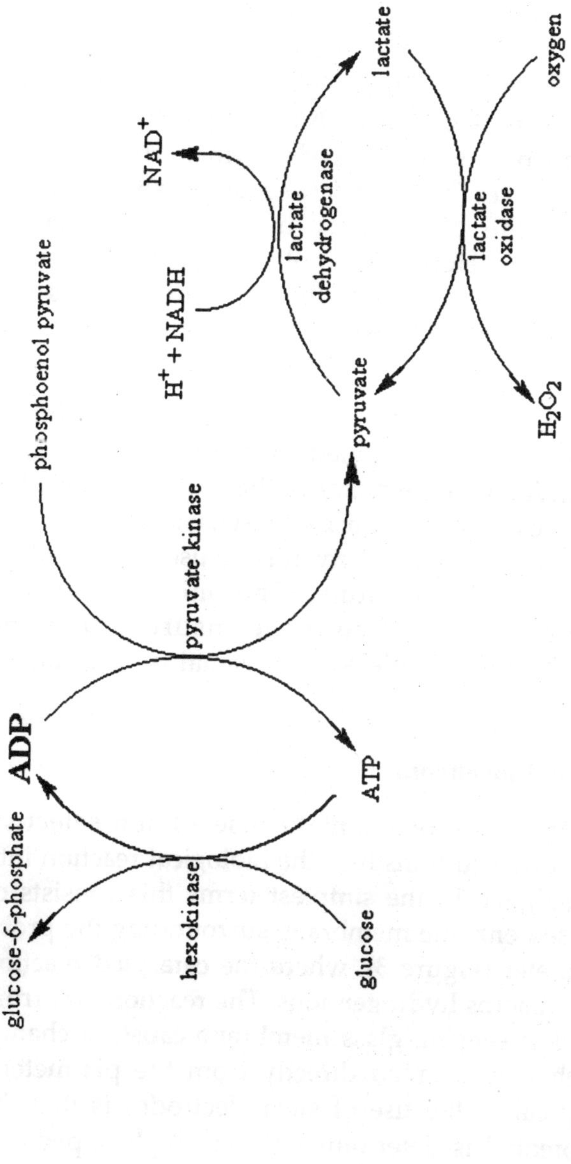

Recycle scheme

oxidase) are co-immobilised within the packed bed reactor. In spite of the positive enthalpy of the pyruvate kinase reaction, the overall process results in a 1000 fold increase in sensitivity, primarily due to the recycling between pyruvate and lactate.

Reaction limitation due to low oxygen solubility may be overcome by replacing it with benzoquinone, which is reduced to hydroquinone by flavo-enzymes. Such reaction systems do, however, have the serious disadvantage in that they increase the probability of the occurrence of interference in the determination of the analyte of interest. Reactions involving the generation of hydrogen ions can be made more sensitive by the inclusion of a base having a high heat of protonation.

For example, the heat output by the penicillinase reaction may be almost doubled by the use of Tris (tris-(hydroxymethyl)aminomethane) as the buffer.In short, the main advantages of the thermistor biosensor are its general applicability and the possibility for its use on turbid or strongly coloured solutions. The most important disadvantage is the difficulty in ensuring that the temperature of the sample stream remains constant (± 0.01°C).

Potentiometric Biosensors

Potentiometric biosensors make use of ion-selective electrodes in order to transduce the biological reaction into an electrical signal. In the simplest terms this consists of an immobilised enzyme membrane surrounding the probe from a pH-meter (Figure 3), where the catalysed reaction generates or absorbs hydrogen ions. The reaction occurring next to the thin sensing glass membrane causes a change in pH which may be read directly from the pH-meter's display. Typical of the use of such electrodes is that the electrical potential is determined at very high impedance allowing effectively zero current flow and causing no interference with the reaction.

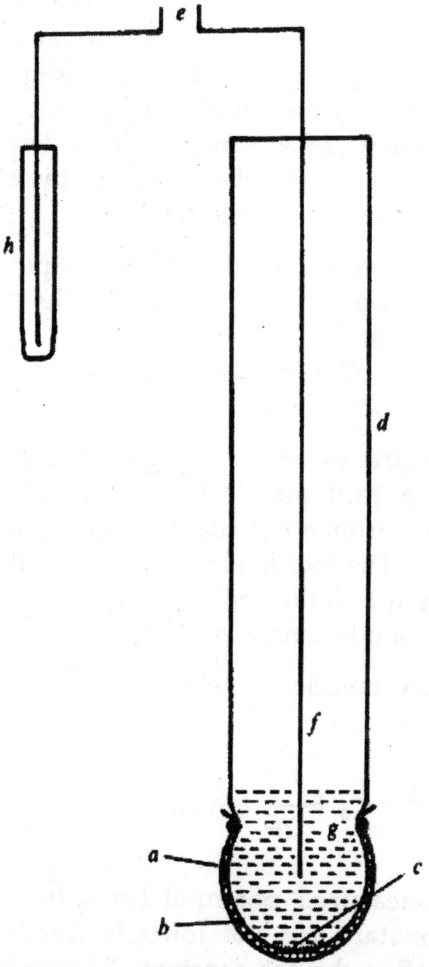

Figure 3. A simple potentiometric biosensor. A semi-permeable membrane (a) surrounds the biocatalyst (b) entrapped next to the active glass membrane (c) of a pH probe (d). The electrical potential (e) is generated between the internal Ag/AgCl electrode (f) bathed in dilute HCl (g) and an external reference electrode (h).

There are three types of ion-selective electrodes which are of use in biosensors:

1. Glass electrodes for cations (e.g. normal pH electrodes) in which the sensing element is a very thin hydrated glass membrane which generates a transverse electrical potential due to the concentration-dependent competition between the cations for specific binding sites. The selectivity of this membrane is determined by the composition of the glass. The sensitivity to H^+ is greater than that achievable for NH_4^+,
2. Glass pH electrodes coated with a gas-permeable membrane selective for CO_2, NH_3 or H_2S. The diffusion of the gas through this membrane causes a change in pH of a sensing solution between the membrane and the electrode which is then determined.
3. Solid-state electrodes where the glass membrane is replaced by a thin membrane of a specific ion conductor made from a mixture of silver sulphide and a silver halide. The iodide electrode is useful for the determination of I- in the peroxidase reaction and also responds to cyanide ions.

The response of an ion-selective electrode is given by

$$E = E_0 + \frac{RT}{zF} Ln([i]) \quad (5)$$

where E is the measured potential (in volts), E_0 is a characteristic constant for the ion-selective/external electrode system, R is the gas constant, T is the absolute temperature (K), z is the signed ionic charge, F is the Faraday, and [i] is the concentration of the free uncomplexed ionic species (strictly, [i] should be the activity of the ion but at the concentrations normally encountered in biosensors, this is effectively equal to the concentration). This means, for example, that there is an increase in the electrical potential of 59 mv for every decade increase in the concentration of H^+ at 25°C.

The logarithmic dependence of the potential on the ionic concentration is responsible both for the wide analytical range and the low accuracy and precision of these sensors. Their normal range of detection is 10^{-4} - 10^{-2} M, although a minority are ten-fold more sensitive. Typical response time are between one and five minutes allowing up to 30 analyses every hour.

Biosensors which involve H^+ release or utilisation necessitate the use of very weakly buffered solutions (i.e. < 5 mM) if a significant change in potential is to be determined. The relationship between pH change and substrate concentration is complex, including other such non-linear effects as pH-activity variation and protein buffering. However, conditions can often be found where there is a linear relationship between the apparent change in pH and the substrate concentration.

A recent development from ion-selective electrodes is the production of ion-selective field effect transistors (ISFETs) and their biosensor use as enzyme-linked field effect transistors (ENFETs, Figure 4). Enzyme membranes are coated on the ion-selective gates of these electronic devices, the biosensor responding to the electrical potential change via the current output.

Thus, these are potentiometric devices although they directly produce changes in the electric current. The main advantage of such devices is their extremely small size (<< 0.1 mm^2) which allows cheap mass-produced fabrication using integrated circuit technology. As an example, a urea-sensitive FET (ENFET containing bound urease with a reference electrode containing bound glycine) has been shown to show only a 15% variation in response to urea (0.05 - 10.0 mg ml^{-1}) during its active lifetime of a month. Several analytes may be determined by miniaturised biosensors containing arrays of ISFETs and ENFETs. The sensitivity of FETs, however, may be affected by the composition, ionic strength and concentrations of the solutions analysed.

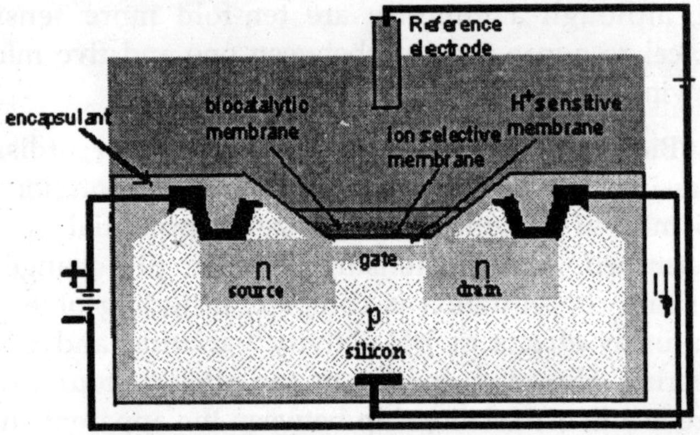

Figure 4. Schematic diagram of the section across the width of an ENFET.

The actual dimensions of the active area is about 500 mm long by 50 mm wide by 300 mm thick. The main body of the biosensor is a p-type silicon chip with two n-type silicon areas; the negative source and the positive drain. The chip is insulated by a thin layer (0.1 mm thick) of silica (SiO_2) which forms the gate of the FET. Above this gate is an equally thin layer of H^+-sensitive material (e.g. tantalum oxide), a protective ion selective membrane, the biocatalyst and the analyte solution, which is separated from sensitive parts of the FET by an inert encapsulating polyimide photopolymer. When a potential is applied between the electrodes, a current flows through the FET dependent upon the positive potential detected at the ion-selective gate and its consequent attraction of electrons into the depletion layer. This current (I) is compared with that from a similar, but non-catalytic ISFET immersed in the same solution.

Amperometric Biosensors

Amperometric biosensors function by the production of a current when a potential is applied between two electrodes. They generally have response times, dynamic ranges and sensitivities similar to the potentiometric biosensors. The simplest amperometric biosensors in common usage involve the Clark oxygen electrode (Figure 5). This consists of a platinum cathode at which oxygen is reduced and a silver/silver chloride reference electrode.

When a potential of -0.6 V, relative to the Ag/AgCl electrode is applied to the platinum cathode, a current proportional to the oxygen concentration is produced. Normally both electrodes are bathed in a solution of saturated potassium chloride and separated from the bulk solution by an oxygen-permeable plastic membrane (e.g. Teflon, polytetrafluoroethylene). The following reactions occur:

$$\text{Ag anode } 4Ag^0 + 4Cl^- \longrightarrow 4AgCl + 4e^-$$

$$\text{Pt cathode } O_2 + 4H^+ + 4e^- \longrightarrow 2H_2O$$

The efficient reduction of oxygen at the surface of the cathode causes the oxygen concentration there to be effectively zero. The rate of this electrochemical reduction therefore depends on the rate of diffusion of the oxygen from the bulk solution, which is dependent on the concentration gradient and hence the bulk oxygen concentration. It is clear that a small, but significant, proportion of the oxygen present in the bulk is consumed by this process; the oxygen electrode measuring the rate of a process which is far from equilibrium, whereas ion-selective electrodes are used close to equilibrium conditions. This causes the oxygen electrode to be much more sensitive to changes in the temperature than potentiometric sensors.

A typical application for this simple type of biosensor is the determination of glucose concentrations by the use of an immobilised glucose oxidase membrane. The reaction results in a reduction of the oxygen concentration as it diffuses through the biocatalytic membrane to the cathode, this being detected by a reduction in the current between the electrodes (Figure 6). Other oxidases may be used in a similar manner for the analysis of their substrates (e.g. alcohol oxidase, D- and L-amino acid oxidases, cholesterol oxidase, galactose oxidase, and urate oxidase)

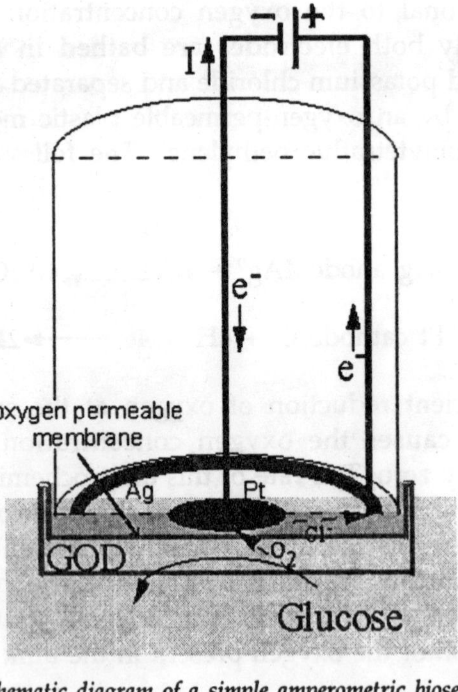

Figure 5. Schematic diagram of a simple amperometric biosensor.

A potential is applied between the central platinum cathode and the annular silver anode. This generates a current (I) which is carried between the electrodes by means of a saturated solution of KCl. This electrode compartment is separated from the biocatalyst (here shown glucose oxidase, GOD) by a thin plastic membrane, permeable only to oxygen.

The analyte solution is separated from the biocatalyst by another membrane, permeable to the substrate(s) and product(s). This biosensor is normally about 1 cm in diameter but has been scaled down to 0.25 mm diameter using a Pt wire cathode within a silver plated steel needle anode and utilising dip-coated membranes.

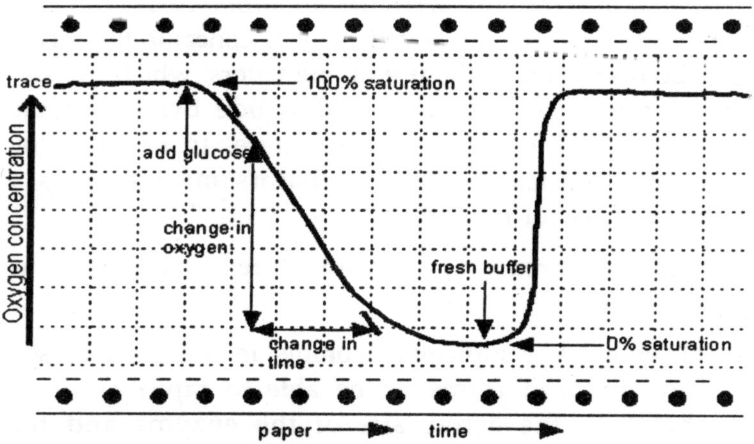

Figure 6. The response of an amperometric biosensor utilising glucose oxidase to the presence of glucose solutions.

Between analyses the biosensor is placed in oxygenated buffer devoid of glucose. The steady rates of oxygen depletion may be used to generate standard response curves and determine unknown samples. The time required for an assay can be considerably reduced if only the initial transient (curved) part of the response need be used, via a suitable model and software. The wash-out time, which roughly equals the time the electrode spends in the sample solution, is also reduced significantly by this process.

An alternative method for determining the rate of this reaction is to measure the production of hydrogen peroxide directly by applying a potential of +0.68 V to the platinum electrode, relative to the Ag/AgCl electrode, and causing the reactions:

$$\text{Pt anode } H_2O_2 \longrightarrow O_2 + 2H^+ + 2e^-$$

$$\text{Ag cathode } 2AgCl + 2e^- \longrightarrow 2Ag^0 + 2Cl^-$$

The major problem with these biosensors is their dependence on the dissolved oxygen concentration. This may be overcome by the use of 'mediators' which transfer the electrons directly to the electrode bypassing the reduction of the oxygen co-substrate. In order to be generally applicable these mediators must possess a number of useful properties.

1. They must react rapidly with the reduced form of the enzyme.
2. They must be sufficiently soluble, in both the oxidised and reduced forms, to be able to rapidly diffuse between the active site of the enzyme and the electrode surface. This solubility should, however, not be so great as to cause significant loss of the mediator from the biosensor's microenvironment to the bulk of the solution. However soluble, the mediator should generally be non-toxic.
3. The overpotential for the regeneration of the oxidised mediator, at the electrode, should be low and independent of pH.

4. The reduced form of the mediator should not readily react with oxygen.

The ferrocenes represent a commonly used family of mediators (Figure 7a). Their reactions may be represented as follows,

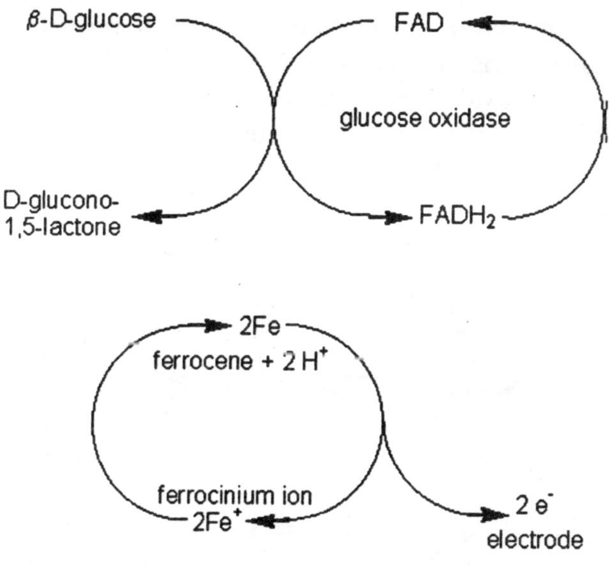

Electrodes have now been developed which can remove the electrons directly from the reduced enzymes, without the necessity for such mediators. They utilise a coating of electrically conducting organic salts, such as N-methylphenazinium cation (NMP^+, Figure 7b) with tetracyanoquinodimethane radical anion (TCNQ.- Figure 7c). Many flavo-enzymes are strongly adsorbed by such organic conductors due to the formation of salt links, utilising the alternate positive and negative charges, within their hydrophobic environment.

Such enzyme electrodes can be prepared by simply dipping the electrode into a solution of the enzyme and they may remain stable for several months. These electrodes can also be used for reactions involving NAD(P)$^+$-dependent dehydrogenases as they also allow the electrochemical oxidation of the reduced forms of these coenzymes.

The three types of amperometric biosensor utilising product, mediator or organic conductors represent the three generations in biosensor development (Figure 8). The reduction in oxidation potential, found when mediators are used, greatly reduces the problem of interference by extraneous material.

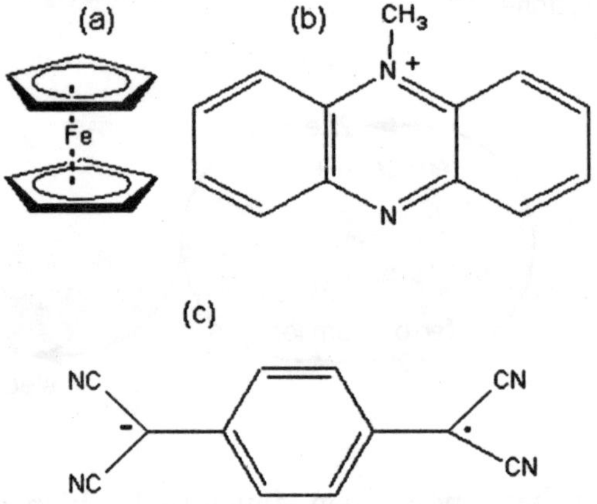

Figure 7. (a) Ferrocene (ε5-bis-cyclopentadienyl iron), the parent compound of a number of mediators. (b) TMP$^+$, the cationic part of conducting organic crystals. (c) TCNQ.-, the anionic part of conducting organic crystals. It is a resonance-stabilised radical formed by the one-electron oxidation of TCNQH$_2$.

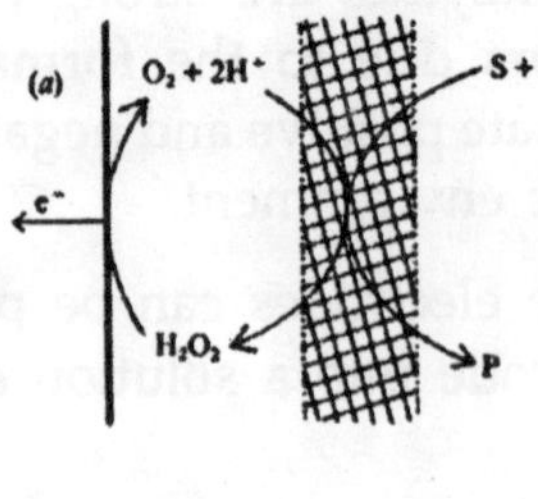

Contd. . .

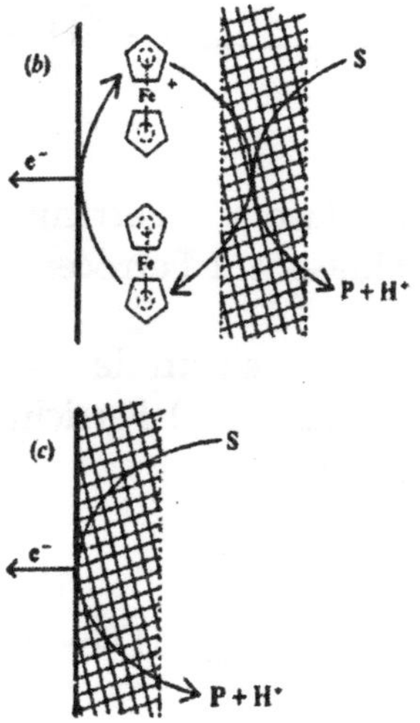

Figure 8. Amperometric biosensors for flavo-oxidase enzymes illustrating the three generations in the development of a biosensor.

The biocatalyst is shown schematically by the cross-hatching. (a) First generation electrode utilising the H_2O_2 produced by the reaction. (E_0 = +0.68 V). (b) Second generation electrode utilising a mediator (ferrocene) to transfer the electrons, produced by the reaction, to the electrode. (E_0 = +0.19 V). (c) Third generation electrode directly utilising the electrons produced by the reaction. (E_0 = +0.10 V). All electrode potentials (E_0) are relative to the $Cl^-/AgCl,Ag^0$ electrode. The following reaction occurs at the enzyme in all three biosensors:

$$\text{Substrate(2H)} + \text{FAD-oxidase} \longrightarrow \text{Product} + \text{FADH}_2\text{-oxidasefi}$$

This is followed by the processes:

(a) biocatalyst

$$\text{FADH}_2\text{-oxidase} + O_2 \longrightarrow \text{FAD-oxidase} + H_2O_2$$

$$H_2O_2 \xrightarrow{\text{electrode}} O_2 + 2H^+ + 2e^-$$

(b) biocatalyst

$$\text{FADH}_2\text{-oxidase} + 2\ \text{Ferricinium}^+ \longrightarrow \text{FAD-oxidase} + 2\ \text{Ferrocene}^+\ 2H^+$$

$$2\ \text{Ferrocene} \xrightarrow{\text{electrode}} 2\ \text{Ferricinium}^+ + 2e\text{-}$$

(c) biocatalyst/electrode

$$\text{FADH}_2\text{-oxidase} \longrightarrow \text{FAD-oxidase} + 2H^+ + 2e^-$$

The current (i) produced by such amperometric biosensors is related to the rate of reaction (v_A) by the expression:

$$i = nFAv_A \tag{6}$$

where n represents the number of electrons transferred, A is the electrode area, and F is the Faraday. Usually the rate of reaction is made diffusionally controlled by use of external membranes. Under these circumstances the electric current produced is proportional to the analyte concentration and independent both of the enzyme and electrochemical kinetics.

Optical Biosensors

There are two main areas of development in optical biosensors. These involve determining changes in light absorption between the reactants and products of a reaction, or measuring the light output by a luminescent process. The former usually involve the widely established, if rather low technology, use of colorimetric test strips.

These are disposable single-use cellulose pads impregnated with enzyme and reagents. The most common use of this technology is for whole-blood monitoring in diabetes control. In this case, the strips include glucose oxidase, horseradish peroxidase (EC 1.11.1.7) and a chromogen (e.g. o-toluidine or 3,3',5,5'-tetramethylbenzidine). The hydrogen peroxide, produced by the aerobic oxidation of glucose, oxidising the weakly coloured chromogen to a highly coloured dye.

$$\text{chromogen(2H)} + H_2O_2 \xrightarrow{\text{peroxidase}} \text{dye} + 2H_2O$$

The evaluation of the dyed strips is best achieved by the use of portable reflectance meters, although direct visual comparison with a coloured chart is often used. A wide variety of test strips involving other enzymes are commercially available at the present time.A most promising biosensor involving luminescence uses firefly luciferase to detect the presence of bacteria in food or clinical samples.

Bacteria are specifically lysed and the ATP released (roughly proportional to the number of bacteria present) reacted with D-luciferin and oxygen in a reaction which produces yellow light in high quantum yield.

$$\text{ATP} + \text{D-luciferin} + O_2 \xrightarrow{\text{luciferase}} \text{oxyluciferin} + \text{AMP} + \text{pyrophosphate} + CO_2 + \text{light (562 nm)}$$

The light produced may be detected photometrically by use of high-voltage, and expensive, photomultiplier tubes or low-voltage cheap photodiode systems. The sensitivity of the photomultiplier-containing systems is, at present, somewhat greater (< 104 cells ml^{-1}, < 10^{-12} M ATP) than the simpler photon detectors which use photodiodes. Firefly luciferase is a very expensive enzyme, only obtainable from the tails of wild fireflies. Use of immobilised luciferase greatly reduces the cost of these analyses.

Piezo-electric Biosensors

Piezo-electric crystals (e.g. quartz) vibrate under the influence of an electric field. The frequency of this oscillation (f) depends on their thickness and cut, each crystal having a characteristic resonant frequency. This resonant frequency changes as molecules adsorb or desorb from the surface of the crystal, obeying the relationshipes

$$\Delta f = \frac{Kf^2 \Delta m}{A} \tag{7}$$

where Δf is the change in resonant frequency (H_z), Dm is the change in mass of adsorbed material (g), K is a constant for the particular crystal dependent on such factors as its density and cut, and A is the adsorbing surface area (cm^2). For any piezo-electric crystal, the change in frequency is proportional to the mass of absorbed material, up to about a 2% change. This frequency change is easily detected by relatively unsophisticated electronic circuits.

A simple use of such a transducer is a formaldehyde biosensor, utilising a formaldehyde dehydrogenase coating immobilised to a quartz crystal and sensitive to gaseous formaldehyde. The major drawback of these devices is the interference from atmospheric humidity and the difficulty in using them for the determination of material in solution. They are, however, inexpensive, small and robust, and capable of giving a rapid response.

Immunosensors

Biosensors may be used in conjunction with enzyme-linked immunosorbent assays (ELISA). The principles behind the ELISA technique is shown in Figure 9. ELISA is used to detect and amplify an antigen-antibody reaction; the amount of enzyme-linked antigen bound to the immobilised antibody being determined by the relative concentration of the free and conjugated antigen and quantified by the rate of enzymic reaction.

Enzymes with high turnover numbers are used in order to achieve rapid response. The sensitivity of such assays may be further enhanced by utilising enzyme-catalysed reactions which give intrinsically greater response; for instance, those giving rise to highly coloured, fluorescent or bioluminescent products. Assay kits using this technique are now available for a vast range of analyses.

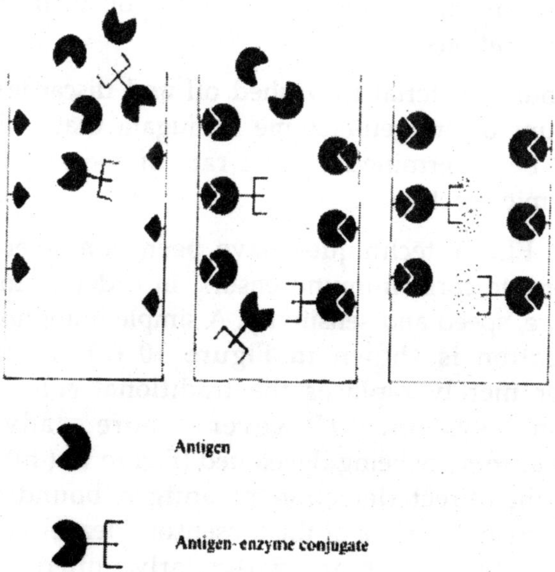

Contd. . .

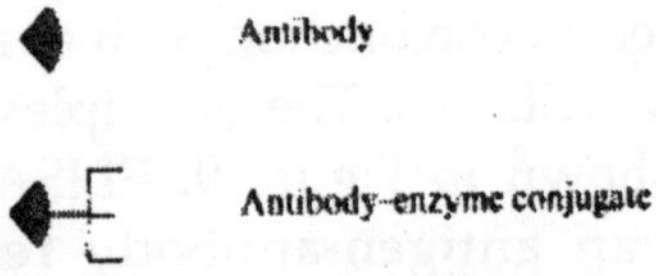

Figure 9. Principles of a direct competitive ELISA.

In the above figure:

(i) Antibody, specific for the antigen of interest is immobilised on the surface of a tube. A mixture of a known amount of antigen-enzyme conjugate plus unknown concentration of sample antigen is placed in the tube and allowed to equilibrate.

(ii) After a suitable period the antigen and antigen-enzyme conjugate will be distributed between the bound and free states dependent upon their relative concentrations.

(iii) Unbound material is washed off and discarded. The amount of antigen-enzyme conjugate that is bound may be determined by the rate of the subsequent enzymic reaction.

Recently ELISA techniques have been combined with biosensors, to form immunosensors, in order to increase their range, speed and sensitivity. A simple immunosensor configuration is shown in Figure 10 (a), where the biosensor merely replaces the traditional colorimetric detection system. However more advanced immunosensors are being developed (Figure 10 (b)) which rely on the direct detection of antigen bound to the antibody-coated surface of the biosensor. Piezoelectric and FET-based biosensors are particularly suited to such applications.

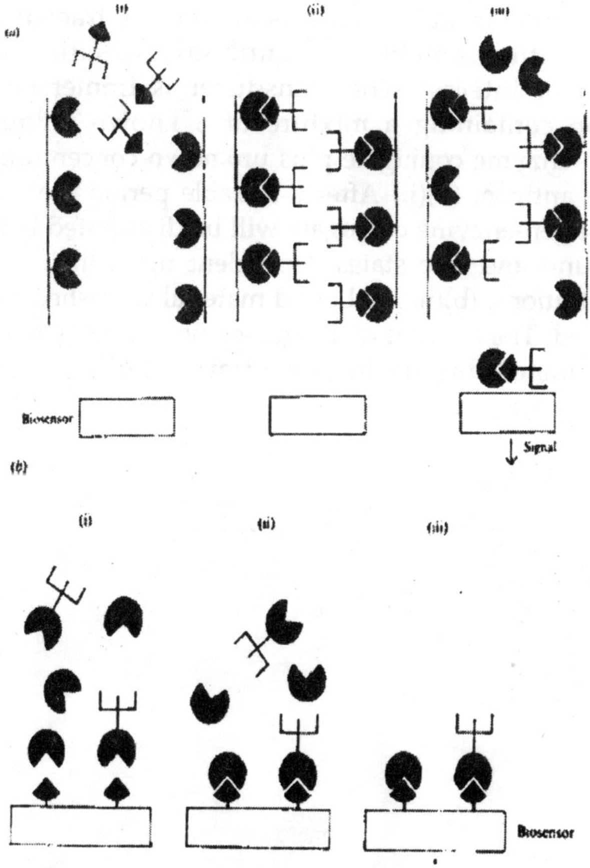

Figure 10. Principles of immunosensors.

(a)(i) A tube is coated with (immobilised) antigen. An excess of specific antibody-enzyme conjugate is placed in the tube and allowed to bind. (a)(ii) After a suitable period any unbound material is washed off. (a)(iii) The analyte antigen solution is passed into the tube, binding and releasing some of the antibody-enzyme conjugate dependent upon the antigen's concentration. The amount of antibody-enzyme conjugate released is determined by the response from the biosensor. (b)(i) A transducer is coated with (immobilised) antibody, specific for the antigen of interest. The transducer is immersed in a solution containing a mixture of a known amount of antigen-enzyme conjugate plus unknown concentration of sample antigen. (b)(ii) After a suitable period the antigen and antigen-enzyme conjugate will be distributed between the bound and free states dependent upon their relative concentrations. (b)(iii) Unbound material is washed off and discarded. The amount of antigen-enzyme conjugate bound is determined directly from the transduced signal.

8

Enzyme Kinetics

The early development of the concept of catalysis in the 19th century went hand in hand with the discovery of powerful catalysts from biological sources. These were called enzymes and were later found to be proteins. They mediate all synthetic and degradative reactions carried out by living organisms. They are very efficient catalysts, often far superior to conventional chemical catalysts, for which reason they are being employed increasingly in today's high-technological society, as a highly significant part of biotechnological expansion. Their utilization has created a billion dollar business including a wide diversity of industrial processes, consumer products, and the burgeoning field of biosensors.

Enzymes have a number of distinct advantages over conventional chemical catalysts. Foremost amongst these are their specificity and selectivity not only for particular reactions but also in their discrimination between similar parts of molecules (regiospecificity) or optical isomers (stereospecificity). They catalyse only the reactions of very narrow ranges of reactants (substrates), which may consist of a small number of closely related classes of compounds (e.g. trypsin catalyses the hydrolysis of some peptides and esters in addition to most proteins), a single class of compounds (e.g. hexokinase catalyses the transfer of a phosphate group from ATP to several hexoses), or a single compound (e.g. glucose oxidase oxidises only glucose amongst the naturally occurring sugars). This means that the chosen reaction can be catalysed to the exclusion of side-reactions, eliminating undesirable by-products. Thus, higher productivities may be achieved, reducing material costs.

As a bonus, the product is generated in an uncontaminated state so reducing purification costs and the downstream environmental burden. Often a smaller number of steps may be required to produce the desired end-product. In addition, certain stereospecific reactions (e.g. the conversion of glucose into fructose) cannot be achieved by classical chemical methods without a large expenditure of time and effort. Enzymes work under generally mild processing conditions of temperature, pressure and pH. This decreases the energy requirements, reduces the capital costs due to corrosion-resistant process equipment and further reduces unwanted side-reactions. The high reaction velocities and straightforward catalytic regulation achieved in enzyme-catalysed reactions allow an increase in productivity with reduced manufacturing costs due to wages and overheads.

There are some disadvantages in the use of enzymes which cannot be ignored but which are currently being addressed and overcome. In particular, the high cost of enzyme isolation and purification still discourages their use, especially in areas which currently have an established alternative procedure. The generally unstable nature of enzymes, when removed from their natural environment, is also a major drawback to their more extensive use.

Enzyme Nomenclature

All enzymes contain a protein backbone. In some enzymes

this is the only component in the structure. However there are additional non-protein moieties usually present which may or may not participate in the catalytic activity of the enzyme. Covalently attached carbohydrate groups are commonly encountered structural features which often have no direct bearing on the catalytic activity, although they may well effect an enzyme's stability and solubility. Other factors often found are metal ions (cofactors) and low molecular weight organic molecules (coenzymes). These may be loosely or tightly bound by noncovalent or covalent forces. They are often important constituents contributing to both the activity and stability of the enzymes. This requirement for cofactors and coenzymes must be recognised if the enzymes are to be used efficiently and is particularly relevant in continuous processes where there may be a tendency for them to become separated from an enzyme's protein moiety.

Enzymes are classified according the report of a Nomenclature Committee appointed by the International Union of Biochemistry. This enzyme commission assigned each enzyme a recommended name and a 4-part distinguishing number. It should be appreciated that some alternative names remain in such common usage that they will be used, where appropriate, in this text. The enzyme commission (EC) numbers divide enzymes into six main groups according to the type of reaction catalysed:

(1) Oxidoreductases which involve redox reactions in which hydrogen or oxygen atoms or electrons are transferred between molecules. This extensive class includes the dehydrogenases (hydride transfer), oxidases (electron transfer to molecular oxygen), oxygenases (oxygen transfer from molecular oxygen) and peroxidases (electron transfer to peroxide). For example: glucose oxidase.

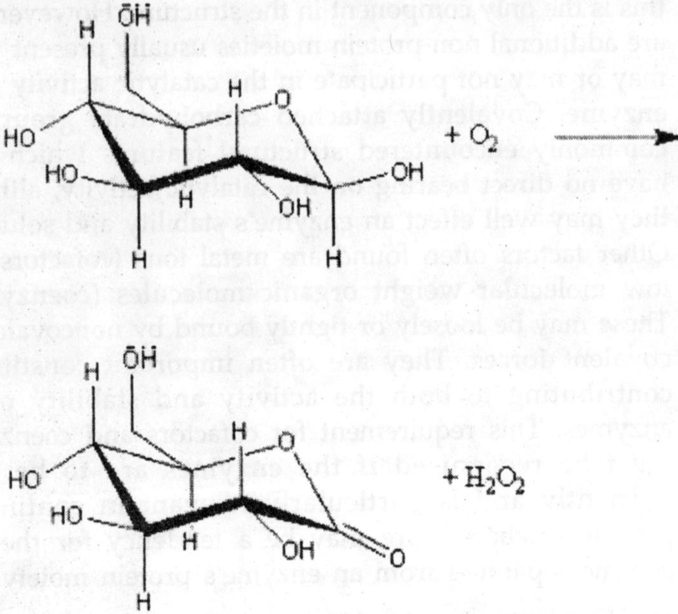

b-D-glucose + oxygen ⟶ D-glucono-1,5-lactone + hydrogen peroxide

(2) Transferases which catalyse the transfer of an atom or group of atoms (e.g. acyl-, alkyl- and glycosyl-), between two molecules, but excluding such transfers as are classified in the other groups (e.g. oxidoreductases and hydrolases).

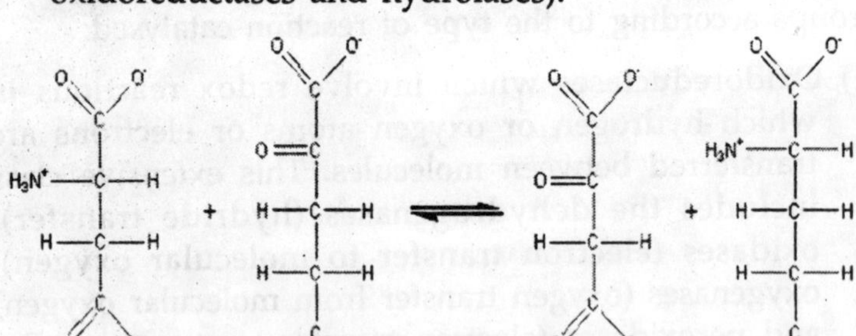

L-aspartate + 2-oxoglutarate ⇌ oxaloacetate + L-glutamate

(3) Hydrolases which involve hydrolytic reactions and their reversal. This is presently the most commonly encountered class of enzymes within the field of enzyme technology and includes the esterases, glycosidases, lipases and proteases. For example: chymosin.

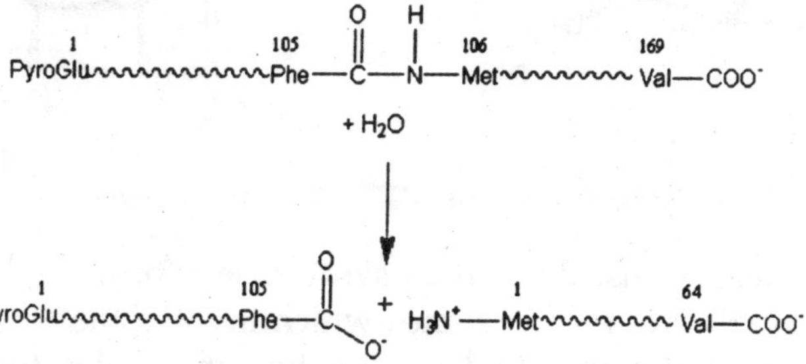

κ-casein + water para κ-casein + caseino macropeptide

(4) Lyases which involve elimination reactions in which a group of atoms is removed from the substrate. This includes the aldolases, decarboxylases, dehydratases and some pectinases but does not include hydrolases. For example: histidine ammonia-lyase.

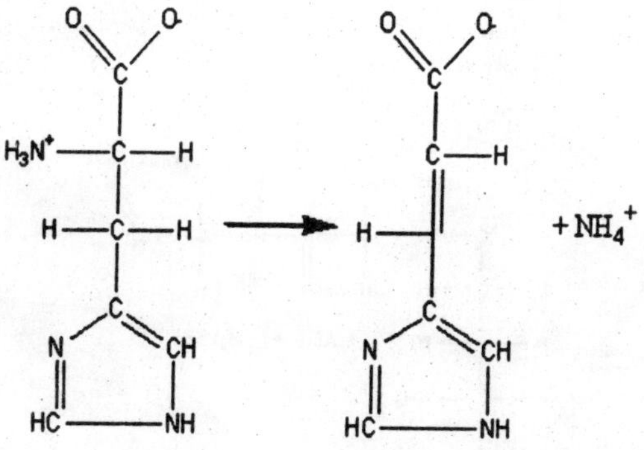

L-histidine ⟶ urocanate + ammonia

(5) Isomerases which catalyse molecular isomerisations and includes the epimerases, racemases and intramolecular transferases. For example: xylose isomerase.

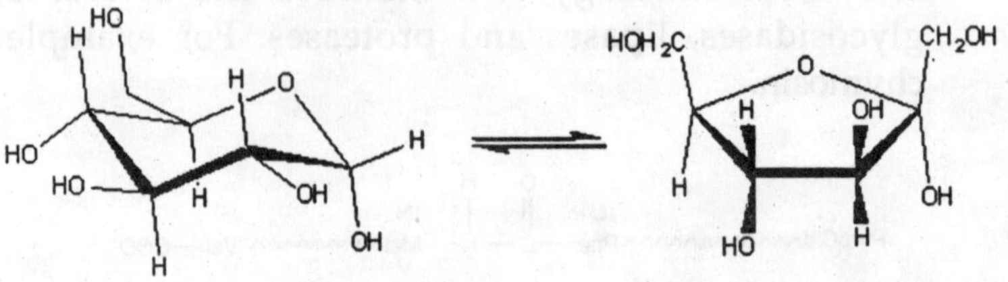

α-D-glucopyranose ⇌ α-D-fructofuranose

(6) Ligases, also known as synthetases, form a relatively small group of enzymes which involve the formation of a covalent bond joining two molecules together, coupled with the hydrolysis of a nucleoside triphosphate. For example: glutathione synthase.

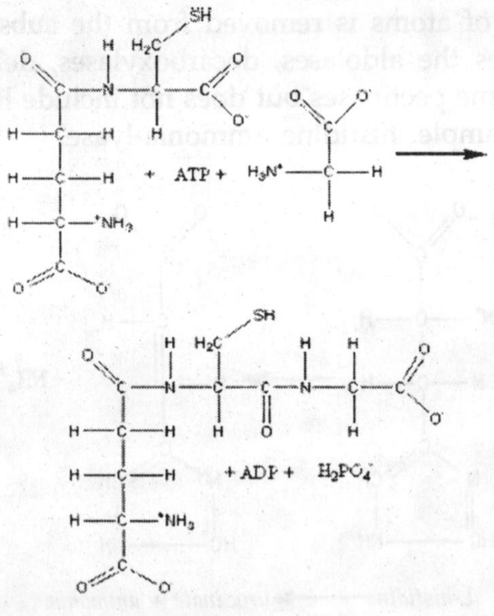

ATP + γ-L-glutamyl-L-cysteine + glycine ADP + phosphate + glutathione

Enzyme Units

The amount of enzyme present or used in a process is difficult to determine in absolute terms (e.g. grams), as its purity is often low and a proportion may be in an inactive, or partially active, state. More relevant parameters are the activity of the enzyme preparation and the activities of any contaminating enzymes. These activities are usually measured in terms of the activity unit (U) which is defined as the amount which will catalyse the transformation of 1 micromole of the substrate per minute under standard conditions.

Typically, this represents 10^{-6} - 10^{-11} Kg for pure enzymes and 10^{-4} - 10^{-7} Kg for industrial enzyme preparations. Another unit of enzyme activity has been recommended. This is the k_{atal} (k_{at}) which is defined as the amount which will catalyse the transformation of one mole of substance per second (1 k_{at} = 60 000 000 U). It is an impracticable unit and has not yet received widespread acceptance.

Sometimes non-standard activity units are used, such as Soxhet, Anson and Kilo Novo units, which are based on physical changes such as lowering viscosity and supposedly better understood by industry. Rightfully, such units are gradually falling into disuse. The activity is a measure of enzyme content that is clearly of major interest when the enzyme is to be used in a process. For this reason, enzymes are usually marketed in terms of activity rather than weight.

The specific activity (e.g. U Kg^{-1}) is a parameter of interest, some utility as an index of purity but lesser importance. There is a major problem with these definitions of activity; the rather vague notion of "standard conditions". These are meant to refer to optimal conditions, especially with regard to pH, ionic strength, temperature, substrate concentration and the presence and concentration of cofactors and coenzymes. However, these so-termed optimal conditions vary both between laboratories and between suppliers. They also depend on the particular application in which the enzyme is to be used.

Additionally, preparations of the same notional specific activity may differ with respect to stability and be capable of very different total catalytic productivity (this is the total substrate converted to product during the lifetime of the catalyst, under specified conditions). Conditions for maximum initial activity are not necessarily those for maximum stability. Great care has to be taken over the consideration of these factors when the most efficient catalyst for a particular purpose is to be chosen

Mechanism of Enzyme Action

In order for a reaction to occur, reactant molecules must contain sufficient energy to cross a potential energy barrier, the activation energy. All molecules possess varying amounts of energy depending, for example, on their recent collision history but, generally, only a few have sufficient energy for reaction. The lower the potential energy barrier to reaction, the more reactants have sufficient energy and, hence, the faster the reaction will occur.

All catalysts, including enzymes, function by forming a transition state, with the reactants, of lower free energy than would be found in the uncatalysed reaction (Figure 1). Even quite modest reductions in this potential energy barrier may produce large increases in the rate of reaction (e.g. the activation energy for the uncatalysed breakdown of hydrogen peroxide to oxygen and water is 76 kJ M^{-1} whereas, in the presence of the enzyme catalase, this is reduced to 30 kJ M^{-1} and the rate of reaction is increased by a factor of 10^8, sufficient to convert a reaction time measured in years into one measured in seconds).

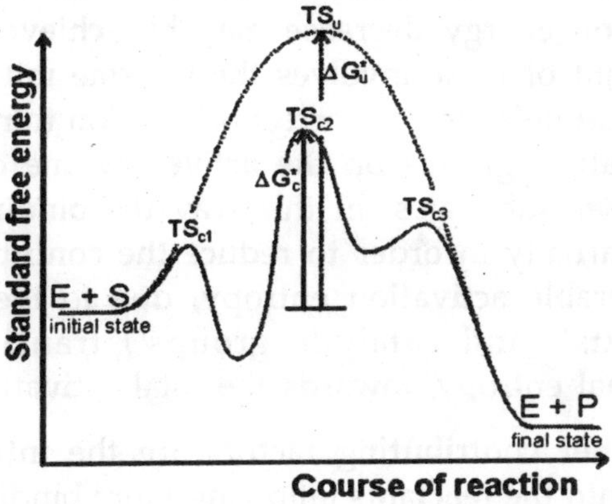

Figure 1. A schematic diagram showing the free energy profile of the course of an enzyme catalysed reaction involving the formation of enzyme-substrate (ES) and enzyme-product (EP) complexes

The catalysed reaction pathway goes through the transition states TS_{c1}, TS_{c2} and TS_{c3}, with standard free energy of activation ΔGc*, whereas the uncatalysed reaction goes through the transition state TSu with standard free energy of activation DGu*. In this example the rate limiting step would be the conversion of ES into EP. Reactions involving several substrates and products, or more intermediates, are even more complicated.

The Michaelis-Menten reaction scheme would give a similar profile but without the EP-complex free energy trough. The schematic profile for the uncatalysed reaction is shown as the dashed line. It should be noted that the catalytic effect only concerns the lowering of the standard free energy of activation from DGu* to DGc* and has no effect on the overall free energy change (i.e.. the difference between the initial and final states) or the related equilibrium constant.

There are a number of mechanisms by which this activation energy decrease may be achieved. The most important of these involves the enzyme initially binding the substrate(s), in the correct orientation to react, close to the catalytic groups on the active enzyme complex and any other substrates. In this way the binding energy is used partially in order to reduce the contribution of the considerable activation entropy, due to the loss of the reactants' (and catalytic groups') translational and rotational entropy, towards the total activation energy.

Other contributing factors are the introduction of strain into the reactants (allowing more binding energy to be available for the transition state), provision of an alternative reactive pathway and the desolvation of reacting and catalysing ionic groups.

The energies available to enzymes for binding their substrates are determined primarily by the complementarity of structures (i.e. a good 3-dimensional fit plus optimal non-covalent ionic and/or hydrogen bonding forces). The specificity depends upon minimal steric repulsion, the absence of unsolvated or unpaired charges, and the presence of sufficient hydrogen bonds. These binding energies are capable of being quite large.

As examples, antibody-antigen dissociation constants are characteristically near 10^{-8} M (free energy of binding is 46 kJ M^{-1}), ATP binds to myosin with a dissociation constant of 10^{-13} M (free energy of binding is 75 kJ M^{-1}) and biotin binds to avidin, a protein found in egg white, with a dissociation constant of 10^{-15} M (free energy of binding is 86 kJ M^{-1}).

However, enzymes do not use this potential binding energy simply in order to bind the substrate(s) and form stable long-lasting complexes. If this were to be the case, the formation of the transition state between ES and EP would involve an extremely large free energy change due to the breaking of these strong binding forces, and the rate of formation of products would be very slow. They must use this binding energy for reducing the free energy of the transition state. This is generally achieved by increasing the binding to the transition state rather than the reactants and, in the process, introducing an energetic strain into the system and allowing more favourable interactions between the enzyme's catalytic groups and the reactants.

Simple Kinetics of Enzyme Action

It is established that enzymes form a bound complex to their reactants (i.e. substrates) during the course of their catalysis and prior to the release of products. This can be simply illustrated, using the mechanism based on that of Michaelis and Menten for a one-substrate reaction, by the reaction sequence:

$$\text{Enzyme} + \text{Substrate} \rightleftharpoons \text{(Enzyme-substrate complex)} \longrightarrow \text{Enzyme} + \text{Product}$$

$$E + S \underset{k_{1}}{\overset{k_{+1}}{\rightleftharpoons}} ES \xrightarrow{k_{+2}} P + E$$

where k_{+1}, k_{-1} and k_{+2} are the respective rate constants, typically having values of 10^5 - 10^8 M^{-1} s^{-1}, 1 - 10^4 s^{-1} and 1 - $10^5 s^{-1}$ respectively; the sign of the subscripts indicating the direction in which the rate constant is acting. For the sake of simplicity the reverse reaction concerning the conversion of product to substrate is not included in this scheme. This is allowable (1) at the beginning of the reaction when there is no, or little, product present, or (2) when the reaction is effectively irreversible. The rate of reaction (v) is the rate at which the product is formed.

$$v = \frac{d[P]}{dt} = k_{+2}[ES] \qquad (1)$$

where [ ] indicates the molar concentration of the material enclosed (i.e. [ES] is the concentration of the enzyme-substrate complex). The rate of change of the concentration of the enzyme-substrate complex equals the rate of its formation minus the rate of its breakdown, forwards to give product or backwards to regenerate substrate.

therefore:

$$\frac{d[ES]}{dt} = k_{+1}[E][S] - (k_{-1} + k_{+2})[ES] \qquad (2)$$

During the course of the reaction, the total enzyme at the beginning of the reaction ($[E]_0$, at zero time) is present either as the free enzyme ([E]) or the ES complex ([ES]).

$$\text{i.e. } [E]_0 = [E] + [ES] \qquad (3)$$

therefore:

$$\frac{d[ES]}{dt} = k_{+1}([E]_0 - [ES])[S] - (k_{-1} + k_{+2})[ES] \qquad (4)$$

Gathering terms together,

$$\frac{d[ES]}{dt} = k_{+1}[E]_0[S] - k_{+1}[ES][S] - (k_{-1} + k_{+2})[ES]$$

$$\frac{d[ES]}{dt} = k_{+1}[E]_0[S] - (k_{+1}[S] + k_{-1} + k_{+2})[ES]$$

this gives:

$$\frac{\frac{d[ES]}{dt}}{k_{+1}[S]+k_{-1}+k_{+2}}+[ES]=\frac{k_{+1}[E]_0[S]}{k_{+1}[S]+k_{-1}+k_{+2}} \tag{5}$$

The differential equation 5 is difficult to handle, but may be greatly simplified if it can be assumed that the left hand side is equal to [ES] alone. This assumption is valid under the sufficient but unnecessarily restrictive steady state approximation that the rate of formation of ES equals its rate of disappearance by product formation and reversion to substrate (i.e. d[ES]/dt is zero). It is additionally valid when the condition:

$$\frac{\frac{d[ES]}{dt}}{k_{+1}[S]+k_{-1}+k_{+2}} \ll [ES] \tag{6}$$

This occurs during a substantial part of the reaction time-course over a wide range of kinetic rate constants and substrate concentrations and at low to moderate enzyme concentrations. The variation in [ES], d[ES]/dt, [S] and [P] with the time-course of the reaction is shown in Figure 2, where it may be seen that the simplified equation is valid throughout most of the reaction.

In figure 2, the simulation shows three distinct phases to the reaction time-course, an initial transient phase which lasts for about a millisecond followed by a longer steady state phase of about 30 minutes when [ES] stays constant but only a small proportion of the substrate reacts. This is followed by the final phase, taking about 6 hours during which the substrate is completely converted to product.

$$\frac{\frac{d[ES]}{dt}}{k_{+1}[S]+k_{-1}+k_{+2}}$$

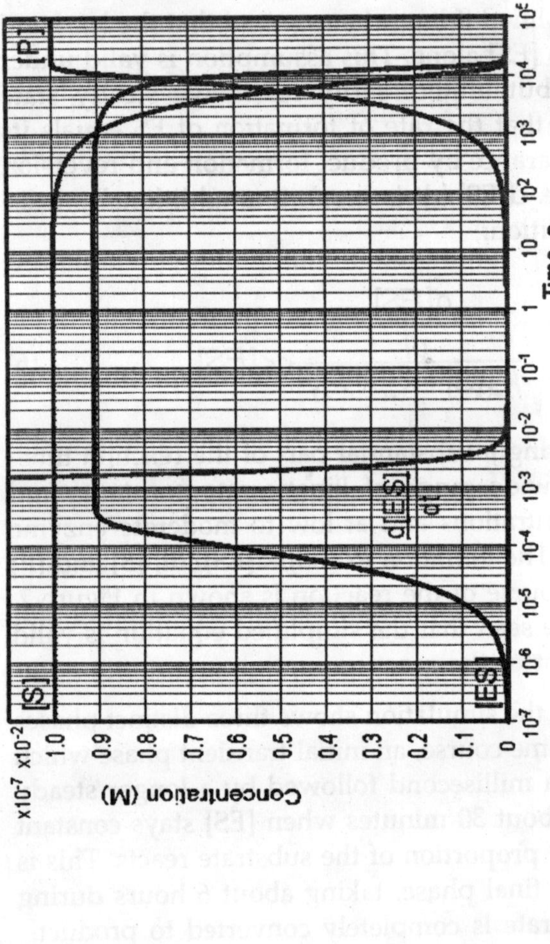

Figure 2. Computer simulation of the progress curves of d[ES]/dt (0 - 10^{-7} M scale), [ES] (0 - 10^{-7} M scale), [S] (0 - 10^{-2} M scale) and [P] (0 - 10^{-2} M scale) for a reaction obeying simple Michaelis-Menten kinetics with k_{+1} = 106 M^{-1} s^{-1}, k^{-1} = 1000 s^{-1}, k_{+2} = 10 s^{-1}, $[E]_0$ = 10^{-7} M and $[S]_0$ = 0.01 M.

is much less than [ES] during both of the latter two phases.

The Michaelis-Menten equation (below) is simply derived from equations 1 and 5, by substituting K_m for

$$\frac{k_{-1}+k_{+2}}{k_{+1}}$$

K_m is known as the Michaelis constant with a value typically in the range 10^{-1} - 10^{-5} M. When $k_{+2} << k_{-1}$, Km equals the dissociation constant (k_{-1}/k_{+1}) of the enzyme substrate complex.

$$v = k_{+2}[ES] = \frac{k_{+2}[E]_0[S]}{[S]+K_m} \qquad (7)$$

or, more simply

$$v = \frac{V_{max}[S]}{[S]+K_m} \qquad (8)$$

where V_{max} is the maximum rate of reaction, which occurs when the enzyme is completely saturated with substrate (i.e. when [S] is very much greater than K_m, V_{max} equals $k_{+2}[E]_0$, as the maximum value [ES] can have is $[E]_0$ when $[E]_0$ is less than $[S]_0$). Equation 8 may be rearranged to show the dependence of the rate of reaction on the ratio of [S] to K_m,

$$v = \frac{V_{max}}{1+\frac{K_m}{[S]}} \qquad (9)$$

and the rectangular hyperbolic nature of the relationship, having asymptotes at v = V_{max} and [S] = $-K_m$,

$$(V_{max}-v)(K_m+[S])=V_{max}K_m \quad (10)$$

The substrate concentration in these equations is the actual concentration at the time and, in a closed system, will only be approximately equal to the initial substrate concentration ($[S]_0$) during the early phase of the reaction. Hence, it is usual to use these equations to relate the initial rate of reaction to the initial, and easily predetermined, substrate concentration (Figure 3). This also avoids any problem that may occur through product inhibition or reaction reversibility.

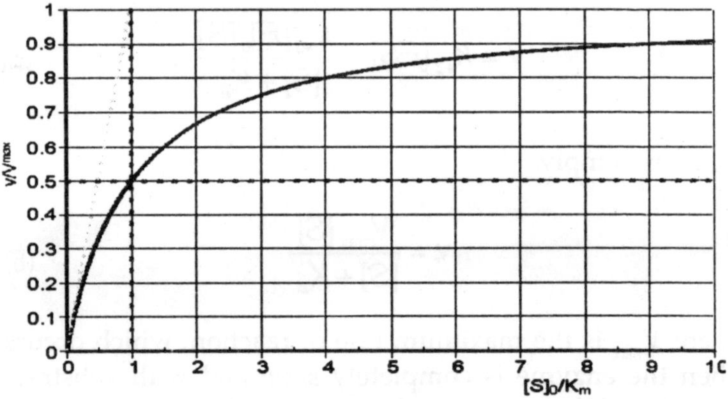

Figure 3. A normalised plot of the initial rate (v_0) against initial substrate concentration ($[S]_0$) for a reaction obeying the Michaelis-Menten kinetics (equation 8). The plot has been normalised in order to make it more generally applicable by plotting the relative initial rate of reaction (v_0/V_{max}) against the initial substrate concentration relative to the Michaelis constant ($[S]_0/K_m$, more commonly referred to as b, the dimensionless substrate concentration).

The curve is a rectangular hyperbola with asymptotes at $v_0 = V_{max}$ and $[S]_0 = -K_m$. The tangent to the curve at the origin goes through the point ($v_0 = V_{max}$),($[S]_0 = K_m$). The ratio V_{max}/K_m is an important kinetic parameter which describes the relative specificity of a fixed amount of the enzyme for its substrate (more precisely defined in terms of k_{cat}/K_m). The substrate concentration, which gives a rate of half the maximum reaction velocity, is equal to the K_m.

It has been established that few enzymes follow the Michaelis-Menten equation over a wide range of experimental conditions. However, it remains by far the most generally applicable equation for describing enzymic reactions. Indeed it can be realistically applied to a number of reactions which have a far more complex mechanism than the one described here.

In these cases K_m remains an important quantity, characteristic of the enzyme and substrate, corresponding to the substrate concentration needed for half the enzyme molecules to bind to the substrate (and, therefore, causing the reaction to proceed at half its maximum rate) but the precise kinetic meaning derived earlier may not hold and may be misleading. In these cases the K_m is likely to equal a much more complex relationship between the many rate constants involved in the reaction scheme.

It remains independent of the enzyme and substrate concentrations and indicates the extent of binding between the enzyme and its substrate for a given substrate concentration, a lower Km indicating a greater extent of binding. V_{max} clearly depends on the enzyme concentration and for some, but not all, enzymes may be largely independent of the specific substrate used.

K_m and V_{max} may both be influenced by the charge and conformation of the protein and substrate(s) which are determined by pH, temperature, ionic strength and other factors. It is often preferable to substitute kcat for k_{+2}, where $V_{max} = k_{cat}[E]_0$, as the precise meaning of k_{+2}, above, may also be misleading. k_{cat} is also known as the turnover number as it represents the maximum number of substrate molecules that the enzyme can 'turn over' to product in a set time (e.g. the turnover numbers of a-amylase, glucoamylase and glucose isomerase are 500 s^{-1}, 160 s^{-1} and 3 s^{-1} respectively; an enzyme with a relative molecular mass of 60000 and specific activity 1 U mg^{-1} has a turnover number of 1 s^{-1}). The ratio k_{cat}/K_m determines the relative rate of reaction at low substrate concentrations, and is known as the specificity constant. It is also the apparent 2nd order rate constant at low substrate concentrations (Figure 3), where

$$v = \frac{k_{cat}}{K_m}[E]_0[S] \tag{11}$$

Many applications of enzymes involve open systems, where the substrate concentration remains constant, due to replenishment, throughout the reaction time-course. This is, of course, the situation that often prevails in vivo. Under these circumstances, the Michaelis-Menten equation is obeyed over an even wider range of enzyme concentrations than allowed in closed systems, and is commonly used to model immobilised enzyme kinetic systems.

Enzymes have evolved by maximising k_{cat}/K_m (i.e. the specificity constant for the substrate) whilst keeping K_m approximately identical to the naturally encountered substrate concentration. This allows the enzyme to operate efficiently and yet exercise some control over the rate of reaction.

The specificity constant is limited by the rate at which the reactants encounter one another under the influence of diffusion. For a single-substrate reaction the rate of encounter between the substrate and enzyme is about 10^8 - 10^9 M^{-1} s^{-1}. The specificity constant of some enzymes approach this value although the range of determined values is very broad.

Effect of pH and Ionic Strength

Enzymes are amphoteric molecules containing a large number of acid and basic groups, mainly situated on their surface. The charges on these groups will vary, according to their acid dissociation constants, with the pH of their environment (Table 1). This will effect the total net charge of the enzymes and the distribution of charge on their exterior surfaces, in addition to the reactivity of the catalytically active groups. These effects are especially important in the neighbourhood of the active sites. Taken together, the changes in charges with pH affect the activity, structural stability and solubility of the enzyme.

Table 1. pKasa and heats of ionisationb of the ionising groups commonly found in enzymes

Group	Usual pK_a range	Approximate charge at pH7	Heats of ionisation (kJ mole[-1])
Carboxyl (C-terminal, glutamic acid, aspartic acid)	3 - 6	-1.0	± 5
Ammonio (N-terminal) (lysine)	7 - 9	+1.0	+45
	9 - 11	+1.0	+45
Imidazolyl (histidine)	5 - 8	+0.5	+30
Guanidyl (arginine)	11 - 13	+1.0	+50
Phenolic (tyrosine)	9 - 12	0.0	+25
Thiol (cysteine)	8 - 11	0.0	+25

There will be a pH, characteristic of each enzyme, at which the net charge on the molecule is zero. This is called the isoelectric point (pI), at which the enzyme generally has minimum solubility in aqueous solutions. In a similar manner to the effect on enzymes, the charge and charge distribution on the substrate(s), product(s) and coenzymes (where applicable) will also be affected by pH changes.

Increasing hydrogen ion concentration will, additionally, increase the successful competition of hydrogen ions for any metal cationic binding sites on the enzyme, reducing the bound metal cation concentration.

Decreasing hydrogen ion concentration, on the other hand, leads to increasing hydroxyl ion concentration which compete against the enzymes' ligands for divalent and trivalent cations causing their conversion to hydroxides and, at high hydroxyl concentrations, their complete removal from the enzyme. The temperature also has a marked effect on ionisations, the extent of which depends on the heats of ionisation of the particular groups concerned (Table 1). The relationship between the change in the pKa and the change in temperature is given by a derivative of the Gibbs-Helmholtz equation:

$$\frac{d(pK_a)}{dT} = \frac{-\Delta H}{2.303RT^2} \tag{12}$$

where T is the absolute temperature (K), R is the gas law constant (8.314 J M^{-1} K^{-1}), DH is the heat of ionisation and the numeric constant (2.303) is the natural logarithm of 10, as pK_a's are based on logarithms with base 10. This variation is sufficient to shift the pI of enzymes by up to one unit towards lower pH on increasing the temperature by 50°C.

These charge variations, plus any consequent structural alterations, may be reflected in changes in the binding of the substrate, the catalytic efficiency and the amount of active enzyme. Both V_{max} and K_m will be affected due to the resultant modifications to the kinetic rate constants k_{+1}, k_{-1} and k_{cat} (k_{+2} in the Michaelis-Menten mechanism), and the variation in the concentration of active enzyme. The effect of pH on the V_{max} of an enzyme catalysed reaction may be explained using the, generally true, assumption that only one charged form of the enzyme is optimally catalytic and therefore the maximum concentration of the enzyme-substrate intermediate cannot be greater than the concentration of this species. In simple terms, assume EH- is the only active form of the enzyme,

$$EH^- + S \underset{k_{-1}}{\overset{k_{+1}}{\rightleftharpoons}} EH^-S \xrightarrow{k_{+2}} P$$

The concentration of EH- is determined by the two dissociations

$$EH_2 \overset{K_{a1}}{\rightleftharpoons} EH^- + H^+$$

$$EH^- \overset{K_{a2}}{\rightleftharpoons} E^{2-} + H^+$$

with

$$K_{a1} = \frac{[EH^-]_0[H^+]}{[EH_2]_0} \tag{13}$$

and

$$K_{a2} = \frac{[E^{2-}]_0[H^+]}{[EH^-]_0} \tag{14}$$

However,

$$[E]_0 = [EH_2]_0 + [EH^-]_0 + [E^{2-}]_0 \tag{15}$$

therefore:

$$[E]_0 = [EH^-]_0\left(\frac{[H^+]}{K_{a1}} + 1 + \frac{K_{a2}}{[H^+]}\right) \tag{16}$$

As the rate of reaction is given by k_{+2}[EH-S] and this is maximal when [EH-S] is maximal (i.e.. when [EH-S] = [EH-]$_0$):

$$V_{max} = k_{+2}[EH^-] = \frac{k_{+2}[E]_0}{\left(\frac{[H^+]}{K_{a1}} + 1 + \frac{K_{a2}}{[H^+]}\right)} \quad (17)$$

The V_{max} will be greatest when

$$\frac{[H^+]}{K_{a1}} = \frac{K_{a2}}{[H^+]} \quad (18)$$

therefore:

$$pH_{optimum} = \frac{pK_{a1} + pK_{a2}}{2} \quad (19)$$

This derivation has involved a number of simplifications on the real situation; it ignores the effect of the ionisation of substrates, products and enzyme-substrate complexes and it assumes EH- is a single ionised species when it may contain a mixture of differently ionised groups but with identical overall charge, although the process of binding substrate will tend to fix the required ionic species.

It does, however, produce a variation of maximum rate with pH which gives the commonly encountered 'bell-shaped' curve (Figure 4). Where the actual reaction scheme is more complex, there may be a more complex relationship between V_{max} and pH. In particular, there may be a change in the rate determining step with pH.

It should be recognised that K_m may change with pH in an independent manner to the V_{max} as it usually involves other, or additional, ionisable groups. It is clear that at lower non-saturating substrate concentrations the activity changes with pH may or may not reflect the changes in V_{max}. It should also be noted from the foregoing discussion that the variation of activity with pH depends on the reaction direction under consideration. The pHoptimum may well be different in the forward direction from that shown by the reverse reaction. This is particularly noticeable when reactions which liberate or utilise protons are considered (e.g. dehydrogenases) where there may well be greater than 2 pH units difference between the pHoptimum shown by the rates of forward and reverse reactions.

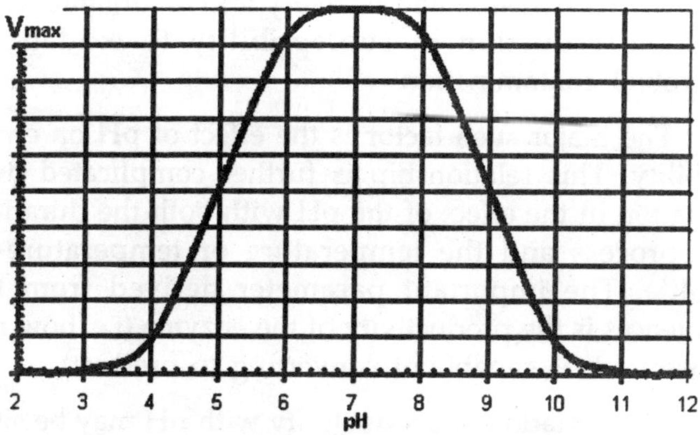

Figure .4. A generally applicable schematic diagram of the variation in the rate of an enzyme catalysed reaction (V_{max}) with the pH of the solution. The centre (optimum pH) and breadth of this 'bell-shaped' curve depend upon the acid dissociation constants of the relevant groups in the enzyme. It should be noted that some enzymes have pH-activity profiles that show little similarity to this diagram.

The variation of activity with pH, within a range of 2-3 units each side of the pI, is normally a reversible process. Extremes of pH will, however, cause a time- and temperature-dependent, essentially irreversible, denaturation. In alkaline solution (pH > 8), there may be partial destruction of cystine residues due to base catalysed β-elimination reactions whereas, in acid solutions (pH < 4), hydrolysis of the labile peptide bonds, sometimes found next to aspartic acid residues, may occur. The importance of the knowledge concerning the variation of activity with pH cannot be over-emphasised.

However, a number of other factors may mean that the optimum pH in the V_{max}-pH diagram may not be the pH of choice in a technological process involving enzymes. These include the variation of solubility of substrate(s) and product(s), changes in the position of equilibrium for a reaction, suppression of the ionisation of a product to facilitate its partition and recovery into an organic solvent, and the reduction in susceptibility to oxidation or microbial contamination.

The major such factor is the effect of pH on enzyme stability. This relationship is further complicated by the variation in the effect of the pH with both the duration of the process and the temperature or temperature-time profile. The important parameter derived from these influences is the productivity of the enzyme (i.e. how much substrate it is capable of converting to product).

The variation of productivity with pH may be similar to that of the V_{max}-pH relationship but changes in the substrate stream composition and contact time may also make some contribution. Generally, the variation must be determined under the industrial process conditions. It is possible to alter the pH-activity profiles of enzymes. The ionisation of the carboxylic acids involves the separation of the released groups of opposite charge.

This process is encouraged within solutions of higher polarity and reduced by less polar solutions. Thus, reducing the dielectric constant of an aqueous solution by the addition of a co-solvent of low polarity (e.g. dioxan, ethanol), or by immobilisation, increases the pK_a of carboxylic acid groups. This method is sometimes useful but not generally applicable to enzyme catalysed reactions as it may cause a drastic change on an enzyme's productivity due to denaturation. The pK_a of basic groups are not similarly affected as there is no separation of charges when basic groups ionise. However, protonated basic groups which are stabilised by neighbouring negatively charged groups will be stabilised (i.e. have lowered pK_a) by solutions of lower polarity.

At higher solution ionic strength, charge separation is encouraged with a concomitant lowering of the carboxylic acid pKas. These changes, extensive as they may be, have little effect on the overall charge on the enzyme molecule at neutral pH and are, therefore, only likely to exert a small influence on the enzyme's isoelectric point. Chemical derivatisation methods are available for converting surface charges from positive to negative and vice-versa.

It is found that a single change in charge has little effect on the pH-activity profile, unless it is at the active site. However if all lysines are converted to carboxylates (e.g. by reaction with succinic anhydride) or if all the carboxylates are converted to amines the profile can be shifted about a pH unit towards higher or lower pH, respectively. The cause of these shifts is primarily the stabilisation or destabilisation of the charges at the active site during the reaction, and the effects are most noticeable at low ionic strength.

The ionic strength of the solution is an important parameter affecting enzyme activity. This is especially noticeable where catalysis depends on the movement of charged molecules relative to each other. Thus both the binding of charged substrates to enzymes and the movement of charged groups within the catalytic 'active' site will be influenced by the ionic composition of the medium. If the rate of the reaction depends upon the approach of charged moieties the following approximate relationship may hold,

$$\log(k) = \log(k_0) + z_A z_B \sqrt{I} \quad (20)$$

where k is the actual rate constant, k0 is the rate constant at zero ionic strength, z_A and z_B are the electrostatic charges on the reacting species, and I is the ionic strength of the solution. If the charges are opposite then there is a decrease in the reaction rate with increasing ionic strength whereas if the charges are identical, an increase in the reaction rate will occur (e.g. the rate controlling step in the catalytic mechanism of chymotrypsin involves the approach of two positively charged groups, ^{57}histidine$^+$ and ^{145}arginine$^+$ causing a significant increase in kcat on increasing the ionic strength of the solution). Even if a more complex relationship between the rate constants and the ionic strength holds, it is clearly important to control the ionic strength of solutions in parallel with the control of pH.

Effect of Temperature and Pressure

Rates of all reactions, including those catalysed by enzymes, rise with increase in temperature in accordance with the Arrhenius equation.

$$k - Ae^{-\Delta G^*/RT} \quad (21)$$

where k is the kinetic rate constant for the reaction, A is the Arrhenius constant, also known as the frequency factor, DG* is the standard free energy of activation (kJ

M^{-1}) which depends on entropic and enthalpic factors, R is the gas law constant and T is the absolute temperature. Typical standard free energies of activation (15 - 70 kJ M^{-1}) give rise to increases in rate by factors between 1.2 and 2.5 for every 10°C rise in temperature. This factor for the increase in the rate of reaction for every 10°C rise in temperature is commonly denoted by the term Q_{10} (i.e. in this case, Q_{10} is within the range 1.2 - 2.5). All the rate constants contributing to the catalytic mechanism will vary independently, causing changes in both K_m and V_{max}.

It follows that, in an exothermic reaction, the reverse reaction (having a higher activation energy) increases more rapidly with temperature than the forward reaction. This, not only alters the equilibrium constant (equation 12), but also reduces the optimum temperature for maximum conversion as the reaction progresses. The reverse holds for endothermic reactions such as that of glucose isomerase where the ratio of fructose to glucose, at equilibrium, increases from 1.00 at 55°C to 1.17 at 80°C.

In general, it would be preferable to use enzymes at high temperatures in order to make use of this increased rate of reaction plus the protection it affords against microbial contamination. Enzymes, however, are proteins and undergo essentially irreversible denaturation (i.e.. conformational alteration entailing a loss of biological activity) at temperatures above those to which they are ordinarily exposed in their natural environment.

These denaturing reactions have standard free energies of activation of about 200 - 300 kJ $mole^{-1}$ (Q_{10} in the range 6 - 36) which means that, above a critical temperature, there is a rapid rate of loss of activity (Figure 5). The actual loss of activity is the product of this rate and the duration of incubation (Figure 6). It may be due to covalent changes such as the deamination of asparagine residues or non-covalent changes such as the rearrangement of the protein chain. Inactivation by heat denaturation has a profound effect on the enzymes productivity (Figure 7).

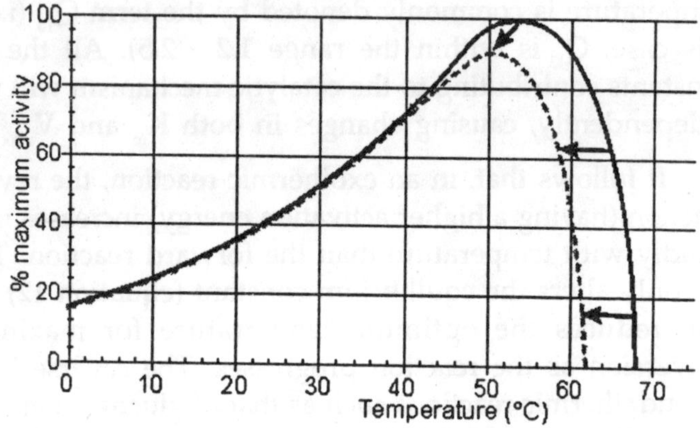

Figure 5. A schematic diagram showing the effect of the temperature on the activity of an enzyme catalysed reaction. — short incubation period; ----- long incubation period. Note that the temperature at which there appears to be maximum activity varies with the incubation time.

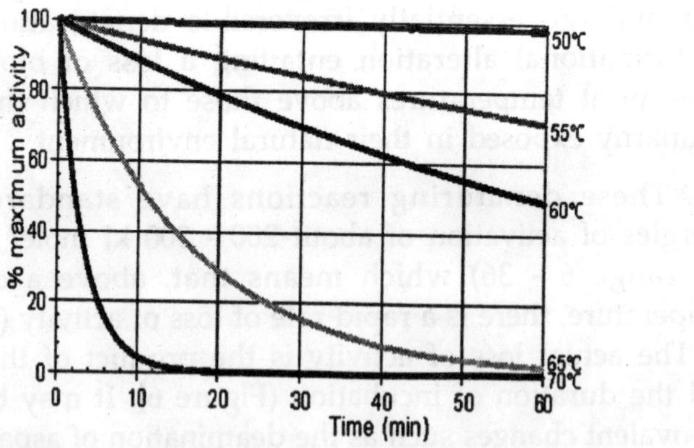

Figure 6. A schematic diagram showing the effect of the temperature on the stability of an enzyme catalysed reaction.

The curves show the percentage activity remaining as the incubation period increases. From the top they represent equal increases in the incubation temperature (50°C, 55°C, 60°C, 65°C and 70°C).

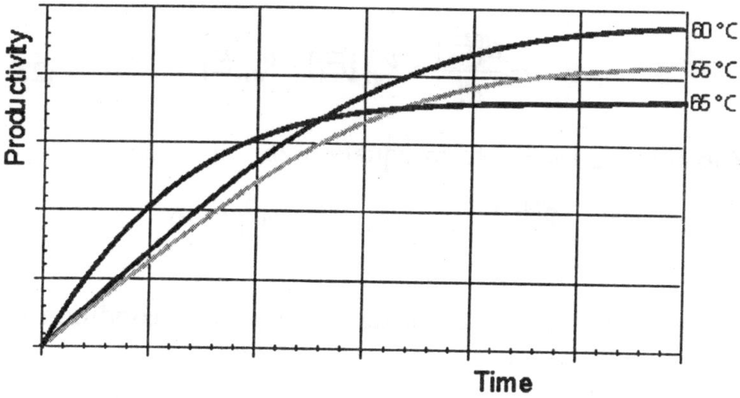

Figure 7. A schematic diagram showing the effect of the temperature on the productivity of an enzyme catalysed reaction.

Assuming, at the beginning of the reaction:

$$[F] = [F]_c \quad (22)$$

and:

$$[E_1] = [E_2] = 0 \quad (23)$$

At time t,

$$[E] + [E_1] + [E_2] = [E]_0 \quad (24)$$

It follows from the reaction scheme,

$$-\frac{d[E]}{dt} = k_{d1}[E] \quad (25)$$

Integrating equation 25 using the boundary condition in equation 22 gives:

$$[E] = [E]_0 e^{(-k_d t)} \qquad (26)$$

From the reaction scheme,

$$-\frac{d[E_1]}{dt} = k_{d2}[E_1] - k_{d1}[E] \qquad (27)$$

Substituting for [E] from equation 26,

$$-\frac{d[E_1]}{dt} = k_{d2}[E_1] - k_{d1}[E]_0 e^{(-k_d t)} \qquad (28)$$

Integrating equation 27 using the boundary condition in equation 23 gives:

$$[E_1] = \frac{k_{d1}[E]_0}{k_{d2} - k_{d1}}\left(e^{(-k_d t)} - e^{(-k_{d2} t)}\right) \qquad (29)$$

If the term 'fractional activity' (Af) is introduced where,

$$A^f = \frac{[E] + A_1[E_1] + A_2[E_2]}{[E]_0} \qquad (30)$$

then, substituting for $[E_2]$ from equation 24, gives:

$$A^f = \frac{[E] + A_1[E_1] + A_2([E]_0 - [E] - [E_1])}{[E]_0} \qquad (31)$$

therefore:

$$A^f = A_2 + \left(1 + \frac{A_1 k_{d1} - A_2 k_{d2}}{k_{d2} - k_{d1}}\right) e^{(-k_{d}t)} -$$

$$\frac{(A_1 - A_2) k_{d1}}{k_{d2} - k_{d1}} e^{(-k_{d2}t)} \quad (32)$$

When both A1 and A2 are zero, the simple first order deactivation rate expression results

$$A^f = e^{(-k_{d}t)} \quad (33)$$

The half-life ($t_{1/2}$) of an enzyme is the time it takes for the activity to reduce to a half of the original activity (i,e. Af = 0.5). If thc enzyme inactivation obeys equation 33, the half-life may be simply derived,

$$Ln(1/2) = -k_{d1} t_{1/2} \quad (34)$$

therefore:

$$t_{1/2} = \frac{0.693}{k_{d1}} \quad (35)$$

In this simple case, the half-life of the enzyme is shown to be inversely proportional to the rate of denaturation.

Many enzyme preparations, both free and immobilised, appear to follow this series-type deactivation scheme. However because reliable and reproducible data is difficult to obtain, inactivation data should, in general, be assumed to be rather error-prone. It is not surprising that such data can be made to fit a model involving four determined parameters (A_1, A_2, k_{d1} and kd_2). Despite this possible reservation, equations 32 and 33 remain quite useful and the theory possesses the definite advantage of simplicity.

In some cases the series-type deactivation may be due to structural microheterogeneity, where the enzyme preparation consists of a heterogeneous mixture of a large number of closely related structural forms. These may have been formed during the past history of the enzyme during preparation and storage due to a number of minor reactions, such as deamidation of one or two asparagine or glutamine residues, limited proteolysis or disulphide interchange. Alternatively it may be due to quaternary structure equilibria or the presence of distinct genetic variants. In any case, the larger the variability the more apparent will be the series-type inactivation kinetics. The practical effect of this is that usually kd1 is apparently much larger than kd2 and A1 is less than unity.

In order to minimise loss of activity on storage, even moderate temperatures should be avoided. Most enzymes are stable for months if refrigerated (0 - 4°C). Cooling below 0°C, in the presence of additives (e.g. glycerol) which prevent freezing, can generally increase this storage stability even further. Freezing enzyme solutions is best avoided as it often causes denaturation due to the stress and pH variation caused by ice-crystal formation.

The first order deactivation constants are often significantly lower in the case of enzyme-substrate, enzyme-inhibitor and enzyme-product complexes which helps to explain the substantial stabilising effects of suitable ligands, especially at concentrations where little free enzyme exists (e.g. [S] >> K_m). Other factors, such as the presence of thiol anti-oxidants, may improve the thermal stability in particular cases.

It has been found that the heat denaturation of enzymes is primarily due to the proteins' interactions with the aqueous environment. They are generally more stable in concentrated, rather than dilute, solutions. In a dry or predominantly dehydrated state, they remain active for considerable periods even at temperatures above 100°C. This property has great technological significance and is currently being exploited by the use of organic solvents.

Pressure changes will also affect enzyme catalysed reactions. Clearly any reaction involving dissolved gases (e.g. oxygenases and decarboxylases) will be particularly affected by the increased gas solubility at high pressures. The equilibrium position of the reaction will also be shifted due to any difference in molar volumes between the reactants and products. However an additional, if rather small, influence is due to the volume changes which occur during enzymic binding and catalysis.

Some enzyme-reactant mixtures may undergo reductions in volume amounting to up to 50 ml mole-1 during reaction due to conformational restrictions and changes in their hydration. This, in turn, may lead to a doubling of the kcat, and/or a halving in the Km for a 1000 fold increase in pressure. The relative effects on kcat and Km depend upon the relative volume changes during binding and the formation of the reaction transition states.

Reversible Reactions

A reversible enzymic reaction (e.g. the conversion of glucose to fructose, catalysed by glucose isomerase) may be represented by the following scheme where the reaction goes through the reversible stages of enzyme-substrate (ES) complex formation, conversion to enzyme-product (EP) complex and finally desorption of the product. No step is completely rate controlling.

$$\mathbf{E + S} \underset{k_{-1}}{\overset{k_{+1}}{\rightleftharpoons}} \mathbf{ES} \underset{k_{-2}}{\overset{k_{+2}}{\rightleftharpoons}} \mathbf{EP} \underset{k_{-3}}{\overset{k_{+3}}{\rightleftharpoons}} \mathbf{E + P}$$

Pairs of symmetrical equations may be obtained for the change in the concentration of the intermediates with time:

$$\frac{d[ES]}{dt} = k_{+1}[E][S] + k_{-2}[EP] - (k_{-1} + k_{+2})[ES] \quad (36)$$

$$\frac{d[EP]}{dt} = k_{-3}[E][P] + k_{+2}[ES] - (k_{+3} + k_{-2})[EP] \quad (37)$$

Assuming that there is no denaturation, the total enzyme concentration must remain constant and:

$$[E] + [ES] + [EP] = [E]_0 \quad (38a)$$

therefore:

$$\frac{d[ES]}{dt} = k_{+1}([E]_0 - [ES] - [EP])[S] + k_{-2}[EP] - (k_{-1} + k_{+2})[ES] \quad (38b)$$

gathering terms in [ES]

$$\frac{d[ES]}{dt} = k_{+1}([E]_0 - [EP])[S] + k_{-2}[EP] - (k_{+1}[S] + k_{-1} + k_{+2})[ES] \quad (39a)$$

$$\frac{\frac{d[ES]}{dt}}{k_{+1}[S] + k_{-1} + k_{+2}} + [ES] = \frac{k_{+1}([E]_0 - [EP])[S] + k_{-2}[EP]}{k_{+1}[S] + k_{-1} + k_{+2}} \quad (39b)$$

and,

$$\frac{d[EP]}{dt} = k_{-3}[([E]_0 - [EP] - [ES])[P] + k_{+2}[ES] - (k_{+3} + k_{-2})[EP] \quad (38c)$$

gathering terms in [EP]

$$\frac{d[EP]}{dt} = k_{-3}[([E]_0 - [ES])[P] + k_{+2}$$

$$[ES] - (k_{-3}[P] + k_{+3} + k_{-2})[EP] \quad (40a)$$

$$\frac{\frac{d[EP]}{dt}}{k_{-3}[P] + k_{+3} + k_{-2}} + [EP] = \frac{k_{-3}([E]_0 - [ES])[P] + k_{+2}[ES]}{k_{-3}[P] + k_{+3} + k_{-2}} \quad (40b)$$

Under similar conditions to those discussed earlier for the Michaelis-Menten mechanism (e.g. under the steady-state assumptions when both d[ES]/dt and d[EP]/dt are zero, or more exactly when

$$\frac{\frac{d[ES]}{dt}}{k_{+1}[S] + k_{-1} + k_{+2}} \langle\langle [ES] \quad (41)$$

and

$$\frac{\frac{d[EP]}{dt}}{k_{-3}[P] + k_{+3} + k_{-2}} \langle\langle [EP] \quad (42)$$

are both true. The following equations may be derived from equation 39b using the approximation, given by equations 41 and collecting terms.

$$k_{+1}([E]_0 - [EP])[S] + k_{-2}[EP] = (k_{+1}[S] + k_{-1} + k_{+2})[ES] \quad (43a)$$

$$-k_{+1}[EP][S] + k_{-2}[EP] = (k_{+1}[S] + k_{-1} + k_{+2})[ES] - \quad (43b)$$

$$[EP] = \frac{(k_{+1}[S] + k_{-1} + k_{+2})[ES] - k_{+1}[E]_0[S]}{k_{-2} - k_{+1}[S]} \quad (43c)$$

Also, the following equations (symmetrical to the above) may be derived from equation 40b by using the approximation, given by equation 42, and collecting terms.

$$k_{-3}([E]_0 - [ES])[P] + k_{+2}[ES] =$$

$$(k_{-3}[P] + k_{+3} + k_{-2})[EP] \quad (1.44a)$$

$$-k_{-3}[ES][P] + k_{+2}[ES] =$$

$$(k_{-3}[P] + k_{+3} + k_{-2})[EP] - k_{-3}[E]_0[P] \quad (44b)$$

$$[ES] = \frac{(k_{-3}[P] + k_{+3} + k_{-2})[EP] - k_{-3}[E]_0[P]}{k_{+2} - k_{-3}[P]} \quad (44c)$$

9

Immobilised Enzymes

Enzyme Reactors

An enzyme reactor consists of a vessel, or series of vessels, used to perform a desired conversion by enzymic means. A number of important types of such reactor are shown diagrammatically in Figure 1. There are several important factors that determine the choice of reactor for a particular process. In general, the choice depends on the cost of a predetermined productivity within the product's specifications. This must be inclusive of the costs associated with substrate(s), downstream processing, labour, depreciation, overheads and process development, in addition to the more obvious costs concerned with building and running the enzyme reactor.

Other contributing factors are the form of the enzyme of choice (i.e. free or immobilised), the kinetics of the reaction and the chemical and physical properties of an immobilisation support including whether it is particulate, membranous or fibrous, and its density, compressibility, robustness, particle size and regenerability. Attention must also be paid to the scale of operation, the possible need for pH and temperature control, the supply and removal of gases and the stability of the enzyme, substrate and product. These factors will be discussed in more detail with respect to the different types of reactor.

Batch reactors generally consist of a tank containing a stirrer (stirred tank reactor, STR). The tank is normally fitted with fixed baffles that improve the stirring efficiency. A batch reactor is one in which all of the product is removed, as rapidly as is practically possible, after a fixed time. Generally this means that the enzyme and substrate molecules have identical residence times within the reactor, although in some circumstances there may be a need for further additions of enzyme and/or substrate (i.e. fed - batch operation). The operating costs of batch reactors are higher than for continuous processes due to the necessity for the reactors to be emptied and refilled both regularly and often. This procedure is not only expensive in itself but means that there are considerable periods when such reactors are not productive; it also makes uneven demands on both labour and services. STRs can be used for processes involving non-immobilised enzymes, if the consequences of these contaminating the product are not severe.

Batch reactors also suffer from pronounced batch-to-batch variations, as the reaction conditions change with time, and may be difficult to scale-up, due to the changing power requirements for efficient fixing. They do, however, have a number of advantageous features. Primary amongst these is their simplicity both in use and in process development. For this reason they are preferred for small-scale production of highly priced products, especially where the same equipment is to be used for a number of different conversions. They offer a closely controllable environment that is useful for slow reactions, where the composition may be accurately monitored, and conditions (e.g. temperature, pH, coenzyme concentrations) varied throughout the reaction. They are also of use when continuous operation of a process proves to be difficult due to the viscous or intractable nature of the reaction mix.

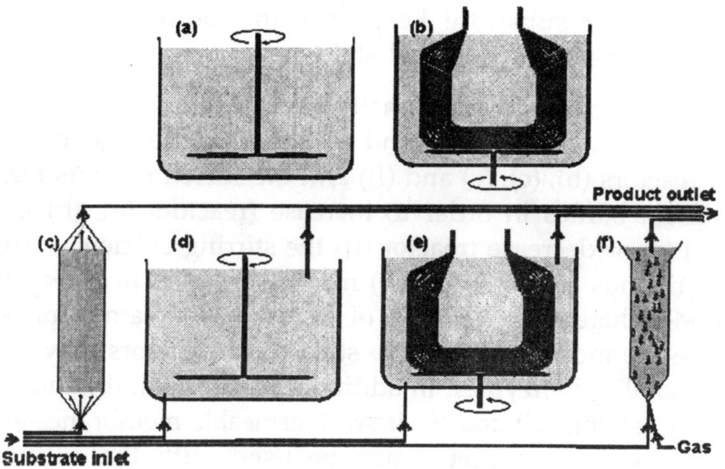

Figure 1. Diagrams of various important enzyme reactor types.

a. Stirred tank batch reactor (STR), which contains all of the enzyme and substrates) until the conversion is complete;

b. batch membrane reactor (MR), where the enzyme is held within membrane tubes which allow the substrate to diffuse in and the product to diffuse out. This reactor may often be used in a semicontinuous manner, using the same enzyme solution for several batches;

c. packed bed reactor (PBR), also called plug -flow reactor (PFR), containing a settled bed of immobilised enzyme particles;

d. continuous flow stirred tank reactor (CSTR) which is a continuously operated version of (a);

e. continuous flow membrane reactor (CMR) which is a continuously operated version of (b);

f. fluidised bed reactor (FBR), where the flow of gas and/or substrate keeps the immobilised enzyme particles in a fluidised state.

All reactors would additionally have heating/cooling coils (interior in reactors (a), and (d), and exterior, generally, in reactors (b), (c), (e) and (f)) and the stirred reactors may contain baffles in order to increase (reactors (a), (b), (d) and (e) or decrease (reactor (f)) the stirring efficiency. The continuous reactors ((c) -(f)) may all be used in a recycle mode where some, or most, of the product stream is mixed with the incoming substrate stream. All reactors may use immobilised enzymes. In addition, reactors (a), (b) and (e) (plus reactors (d) and (f), if semipermeable membranes are used on their outlets) may be used with the soluble enzyme.

The expected productivity of a batch reactor may be calculated by, assuming the validity of the non -reversible Michaelis -Menten reaction scheme with no diffusional control, inhibition or denaturation and equation. The rate of reaction (v) may be expressed in terms of the volume of substrate solution within the reactor (VolS) and the time (t):

$$v = -Vol_S \frac{d[S]}{dt} = \frac{V_{max}[S]}{K_m + [S]} \qquad (1)$$

Therefore:

$$\int_0^t \frac{V_{max}}{Vol_S} dt = -\int_{[S]_0}^{[S]_t} \left(1 + \frac{K_m}{[S]}\right) d[S] \qquad (2)$$

On integrating using the boundary condition that [S] = [S]0 at time (t) = 0:

$$\frac{V_{max}}{Vol_s}t = ([S]_0 - [S]) - K_m \ln\left(\frac{[S]}{[S]_0}\right) \tag{3}$$

Let the fractional conversion be X, where:

$$X = \frac{[S]_0 - [S]}{[S]_0} \tag{4}$$

Therefore;

$$X = 1 - \left(\frac{[S]}{[S]_0}\right) \tag{4a}$$

and

$$\left(\frac{[S]}{[S]_0}\right) = 1 - X \tag{4b}$$

Also

$$[S]_0 - [S] = [S]_0 X \tag{4c}$$

Therefore substituting using (4c) and (4b) in (3):

$$\frac{V_{max}}{Vol_s}t = [S]_0 X - K_m \ln(1 - X) \tag{5}$$

The change in fractional conversion and concentrations of substrate and product with time in a batch reactor is shown in Figure 2(a).

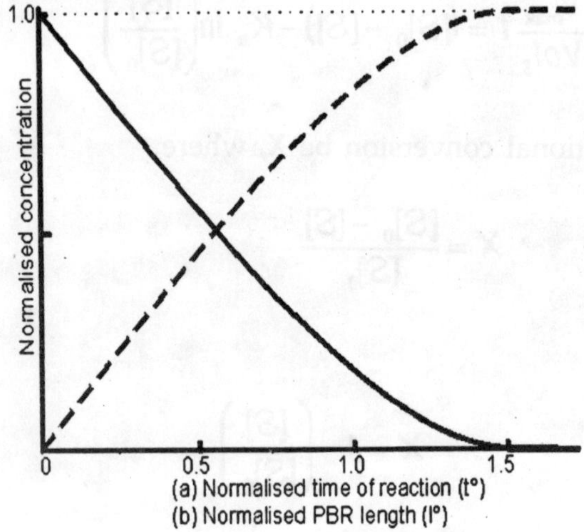

Figure 2. This figure shows two related behaviours.

(a) The change in substrate and product concentrations with time, in a batch reactor. The reaction S——►P is assumed, with the initial condition $[S]_0/K_m$ = 10. The concentrations of substrate (—— and product (----------) are both normalised with respect to $[S]_0$. The normalised time (i.e. t° = t $V_{max}/[S]_0$) is relative to the time (t° = 1) that would be required to convert all the substrate if the enzyme acted at V_{max} throughout, the actual time for complete conversion being longer due to the reduction in the substrate concentration at the reaction progresses. The dashed line also indicates the variation of the fractional conversion (X) with t°.

(b) The change in substrate and product concentrations with reactor length for a PBR. The reaction S——►P is assumed with the initial condition, $[S]_0/K_m$ = 10. The concentrations of substrate (——) and product (-----------) are both normalised with respect to [S]". The normalised reactor length (i.e. I° = lVmax/F, where Vmax is the maximum velocity for unit reactor length and I is the reactor length) is relative to the length (i.e. when I° = 1) that contains sufficient enzyme to convert all the substrate at the given flow rate if the enzyme acted at its maximum velocity throughout; the actual reactor length necessary for complete conversion being longer due to the reduction in the substrate concentration as the reaction progresses. P may be considered as the relative position within a PBR or the reactor's absolute length.

Membrane Reactors

The main requirement for a membrane reactor (MR) is a semipermeable membrane which allows the free passage of the product molecules but contains the enzyme molecules. A cheap example of such a membrane is the dialysis membrane used for removing low molecular weight species from protein preparations. The usual choice for a membrane reactor is a hollow-fibre reactor consisting of a preformed module containing hundreds of thin tubular fibres each having a diameter of about 200 μm and a membrane thickness of about 50 μm.

Membrane reactors may be used in either batch or continuous mode and allow the easy separation of the enzyme from the product. They are normally used with soluble enzymes, avoiding the costs and problems associated with other methods of immobilisation and some of the diffusion limitations of immobilised enzymes. If the substrate is able to diffuse through the membrane, it may be introduced to either side of the membrane with respect to the enzyme, otherwise it must be within the same compartment as the enzyme, a configuration that imposes a severe restriction on the flow rate through the reactor, if used in continuous mode.

Due to the ease with which membrane reactor systems may be established, they are often used for production on a small scale (g to kg), especially where a multi-enzyme pathway or coenzyme regeneration is needed. They allow the easy replacement of the enzyme in processes involving particularly labile enzymes and can also be used for biphasic reactions. The major disadvantage of these reactors concerns the cost of the membranes and their need to be replaced at regular intervals.

The kinetics of membrane reactors are similar to those of the batch STR, in batch mode, or the CSTR, in continuous mode. Deviations from these models occur primarily in configurations where the substrate stream is on the side of the membrane opposite to the enzyme and the reaction is severely limited by its diffusion through the membrane and the products' diffusion in the reverse direction.

Under these circumstances the reaction may be even more severely affected by product inhibition or the limitations of reversibility than is indicated by these models.

Continuous Flow Reactors

The advantages of immobilised enzymes as processing catalysts are most markedly appreciated in continuous flow reactors. In these, the average residence time of the substrate molecules within the reactor is far shorter than that of the immobilised-enzyme catalyst. This results in a far greater productivity from a fixed amount of enzyme than is achieved in batch processes. It also allows the reactor to handle substrates of low solubility by permitting the use of large volumes containing low concentrations of substrate.

The constant reaction conditions may be expected to result in a purer and more reproducible product. There are two extremes of process kinetics in relation to continuous flow reactors;

a. the ideal continuous flow stirred tank reactor (CSTR), in which the reacting stream is completely and rapidly mixed with the whole of the reactor contents and the enzyme contacts low substrate and high product concentrations;. and

b. the ideal continuously operated packed bed reactor (PBR), where no mixing takes place and the enzyme contacts high substrate and low product concentrations.

The properties of the continuously operated fluidised bed reactor (FBR) lie, generally, somewhere between these extremes. An ordered series of CSTRs or FBRs may approximate, in use where the outlet of one reactor forms the inlet to the next reactor, to an equivalent PBR.

The transport of momentum, heat and mass in these continuous reactors are important factors contributing to the resultant productivity. They are due to fluid flow and molecular and turbulent motions, and often described by means of a number of empirical relationships, involving dimensionless numbers. The most important of these is the Reynolds number (*Re*), which relates the inertial force due to the flow of solution to the viscous force resisting that flow.

Low *Re* indicates streamlined flow whereas higher Re indicates progressively more turbulence, there being a critical value for *Re*, dependent on the configuration of the system, at which there is a transition from streamlined flow to turbulent flow. Re is defined in terms of L_{fm}/η or L_f/ν, where L is the characteristic length of the system, fm is the mass flow rate (g m^{-2} s^{-1}), f is the fluid velocity (m s^{-1}), h is the dynamic viscosity (g m^{-1} s^{-1}) and n is the kinematic viscosity (m^2 s^{-1}). Examples of the characteristic lengths in these definitions of Re are the particle diameter for flow past spherical particles, and the tube diameter for flow within hollow tubes. The value of the (characteristic length of the system) x (fluid velocity) for a stirred tank system may be taken as the (stirrer speed) x (stirrer diameter)^2. *Re* is, therefore, higher at high flow rates and low viscosities. Another dimensionless number which is of use in describing and comparing continuous enzyme reactors is the Le Goff number (Lf), which expresses the efficiency with which the energy dissipated in producing the fluid flow is used to transport material (and heat) to the catalytic surface.

Low values for *Lf* indicate a relatively high energetic and financial cost, in achieving contact between the catalytically active immobilised-enzyme surface and the substrate stream, and the consequent reduction in any external diffusion limitations. The *Lf* is higher for low pressure drops through the reactor, high flow rates and high conversions. The relationship between the *Lf* and the Re is shown schematically for a PBR and a CSTR in Figure 3, where it may be seen that the PBR is more efficient at fairly low *Re* but far less so at higher *Re*, reflecting the necessity for stirring the CSTR even at low flow rates and the increased backpressure in PBRs at high flow rates.

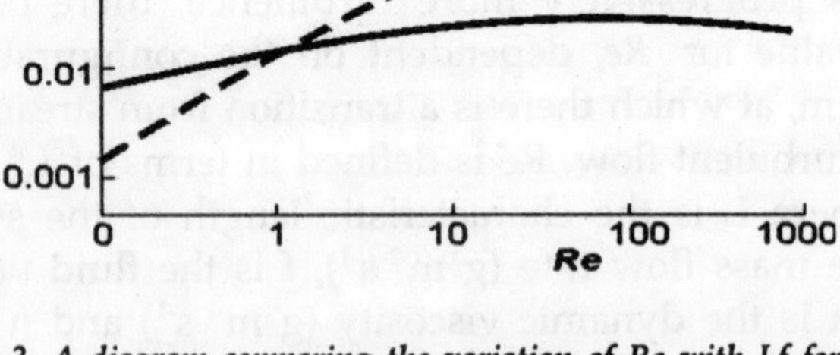

Figure 3. A diagram comparing the variation of Re with Lf for a PBR (——) and a CSTR (-----------). An FBR behaves similarly to the CSTR.

Packed Bed Reactors

The most important characteristic of a PBR is that material flows through the reactor as a plug; they are also called plug flow reactors (PFR). Ideally, all of the substrate stream flows at the same velocity, parallel to the reactor axis with no back -mixing. All material present at any given reactor cross -section has had an identical residence time.

The longitudinal position within the PBR is, therefore, proportional to the time spent within the reactor; all product emerging with the same residence time and all substrate molecule having an equal opportunity for reaction. The conversion efficiency of a PBR, with respect to its length, behaves in a manner similar to that of a well -stirred batch reactor with respect to its reaction time (Figure 2(b)) Each volume element behaves as a batch reactor as it passes through the PBR. Any required degree of reaction may be achieved by use of an idea PBR of suitable length.

The flow rate (F) is equivalent to VolS/t for a batch reactor. Therefore equation (5) may be converted to represent an ideal PBR, given the assumption, not often realised in practice, that there are no diffusion limitations:

$$\frac{V_{max}}{F} = [S]_0 X - K_m \ln(1 - X) \tag{6}$$

In order to produce ideal plug -flow within PBRs, a turbulent flow regime is preferred to laminar flow, as this causes improved mixing and heat transfer normal to the flow and reduced axial back-mixing. Achievement of high enough *Re* may, however, be difficult due to unacceptably high feed rates. Consequent upon the plug -flow characteristic of the PBR is that the substrate concentration is maximised, and the product concentration minimised, relative to the final conversion at every point within the reactor; the effectiveness factor being high on entry to the reactor and low close to the exit. This means that PBRs are the preferred reactors, all other factors being equal, for processes involving product inhibition, substrate activation and reaction reversibility.

At low Re the flow rate is proportional to the pressure drop across the PBR. This pressure drop is, in turn, generally found to be proportional to the bed height, the linear flow rate and dynamic viscosity of the substrate stream and $(1 - \varepsilon)^2/\varepsilon^3$ (where e is the porosity of the reactor; i.e. the fraction of the PBR volume taken up by the liquid phase), but inversely proportional to the cross-sectional area of the immobilised enzyme pellets.

In general PBRs are used with fairly rigid immobilised-enzyme catalysts (1-3 mm diameter), because excessive increases in this flow rate may distort compressible or physically weak particles. Particle deformation results in reduced catalytic surface area of particles contacting the substrate-containing solution, poor external mass transfer characteristics and a restriction to the flow, causing increased pressure drop. A vicious circle of increased back-pressure, particle deformation and restricted flow may eventually result in no flow at all through the PBR.

PBRs behave as deep-bed filters with respect to the substrate stream. It is necessary to use a guard bed if plugging of the reactor by small particles is more rapid than the biocatalysts' deactivation. They are also easily fouled by colloidal or precipitating material. The design of PBRs does not allow for control of pH, by addition of acids or bases, or for easy temperature control where there is excessive heat output, a problem that may be particularly noticeable in wide reactors (> 15 cm diameter).

Deviations from ideal plug-flow are due to back-mixing within the reactors, the resulting product streams having a distribution of residence times. In an extreme case, back-mixing may result in the kinetic behaviour of the reactor approximating to that of the CSTR, and the consequent difficulty in achieving a high degree of conversion. These deviations are caused by channelling, where some substrate passes through the reactor more rapidly, and hold-up, which involves stagnant areas with negligible flow rate.

Channels may form in the reactor bed due to excessive pressure drop, irregular packing or uneven application of the substrate stream, causing flow rate differences across the bed. The use of a uniformly sized catalyst in a reactor with an upwardly flowing substrate stream reduces the chance and severity of non-ideal behaviour.

Continuous Flow Stirred Tank Reactors

This reactor consists of a well-stirred tank containing the enzyme, which is normally immobilised. The substrate stream is continuously pumped into the reactor at the same time as the product stream is removed. If the reactor is behaving in an ideal manner, there is total back-mixing and the product stream is identical with the liquid phase within the reactor and invariant with respect to time.

Some molecules of substrate may be removed rapidly from the reactor, whereas others may remain for substantial periods. The distribution of residence times for molecules in the substrate stream is shown in Figure 4.

The CSTR is an easily constructed, versatile and cheap reactor, which allows simple catalyst charging and replacement. Its well-mixed nature permits straightforward control over the temperature and pH of the reaction and the supply or removal of gases. CSTRs tend to be rather large as the: need to be efficiently mixed. Their volumes are usually about five to ten time the volume of the contained immobilised enzyme. This, however, has the advantage that there is very little resistance to the flow of the substrate stream, which may contain colloidal or insoluble substrates, so long as the insoluble particles are not able to sweep the immobilised enzyme from the reactor.

The mechanical nature of the stirring limits the supports for the immobilised enzymes to materials which do not easily disintegrate to give 'fines' which may enter the product stream. However, fairly small particle (down to about 10 mm diameter) may be used, if they are sufficiently dense to stay within the reactor. This minimises problems due to diffusional resistance.

An ideal CSTR has complete back -mixing resulting in a minimisation of the substrate concentration, and a maximisation of the product concentration, relative to the final conversion, at every point within the reactor the effectiveness factor being uniform throughout. Thus, CSTRs are the preferred reactors, everything else being equal, for processes involving substrate inhibition or product activation. They are also useful where the substrate stream contains an enzyme inhibitor, as it is diluted within the reactor.

This effect is most noticeable if the inhibitor concentration is greater than the inhibition constant and $[S]_0/K_m$ is low for competitive inhibition or high for uncompetitive inhibition, when the inhibitor dilution has more effect than the substrate dilution. Deviations from ideal CSTR behaviour occur when there is a less effective mixing regime and may generally be overcome by increasing the stirrer speed, decreasing the solution viscosity or biocatalyst concentration or by more effective reactor baffling.

The rate of reaction within a CSTR can be derived from a simple mass balance to be the flow rate (F) times the difference in substrate concentration between the reactor inlet and outlet. Hence:

$$F([S]_0 - [S]) = \frac{V_{max}[S]}{K_m + [S]} \tag{7}$$

Therefore:

$$\frac{V_{max}}{F} = K_m \frac{([S]_0 - [S])}{[S]} + ([S]_0 - [S]) \tag{8}$$

from equation (4):

$$\frac{[S]_0 - [S]}{[S]} = \frac{X}{1-X} \tag{9}$$

Therefore:

$$\frac{V_{max}}{F} = [S]_0 X + K_m \frac{X}{1-X} \tag{10}$$

This equation should be compared with that for the PBR (equation (6)). Together these equations can be used for comparing the productivities of the two reactors (Figures 5 and 6).

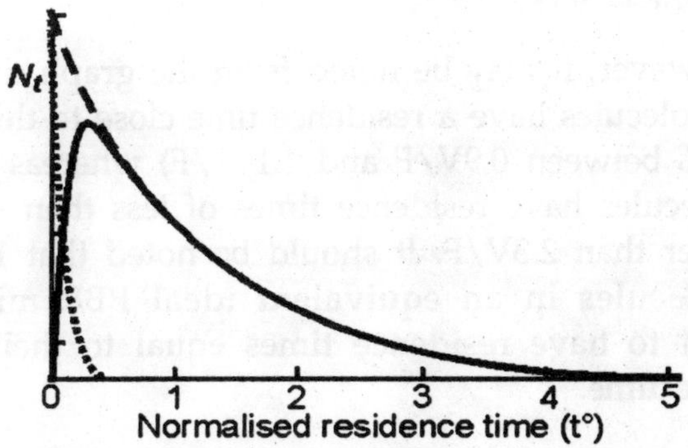

Figure 4. The residence time distribution of a CSTR.

The relative number of molecules resident within the reactor for a particular time N, is plotted against the normalised residence time (i.e. tF/V, where V is the reactor volume, and F is the flow rate; it is the time relative to that required for one reactor volume to pass through the reactor).

The residence time distribution of non -reacting media molecules (----------- which obeys the relationship $[M_t] = [M_0]e^{\frac{-tF}{V}}$ where [M] is the concentration of media molecules, giving a half-life for remaining in the reactor of $Ln(2)x\frac{V}{f} = 0.7\frac{V}{f}$, product (——) and substrate (·········) are shown. The reaction S——►P is assumed, and substrate molecules that have long residence times are converted into product, the average residence time of the product being greater than that for the substrate. The composition of the product stream is identical with that of the liquid phase within the reactor.

This composition may be calculated from the relative areas under the curves and, in this case, represents a 90% conversion. Under continuous operating conditions (operating time > 4V/F), the mean residence time within the reactor is V/F.

However, it may be noted from the graph that only a few molecules have a residence time close to this value (only 7% between 0.9V/F and 1.1 V/F) whereas 20% of the molecules have residence times of less than 0.1 V/F or greater than 2.3V/F. It should be noted that 100% of the molecules in an equivalent ideal PBR might be expected to have residence times equal to their mean residence time.

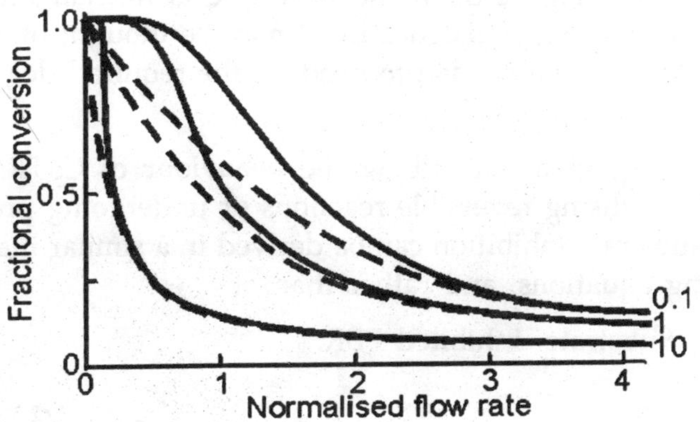

Figure 5. Comparison of the changes in fractional conversion with flow rate between the PBR (——) and CSTR (-----------) at different values of $[S]_0/K_m$ (10,1 and 0.1, higher $[S]_0/K_m$ giving the higher curves). The flow rate is normalised with respect to the reactor's volumetric enzyme content (= FK_m/V_{max}. It can be seen that there is little difference between the two reactors at faster flow rates and lower conversions, especially at high values of $[S]_0/K_m$.

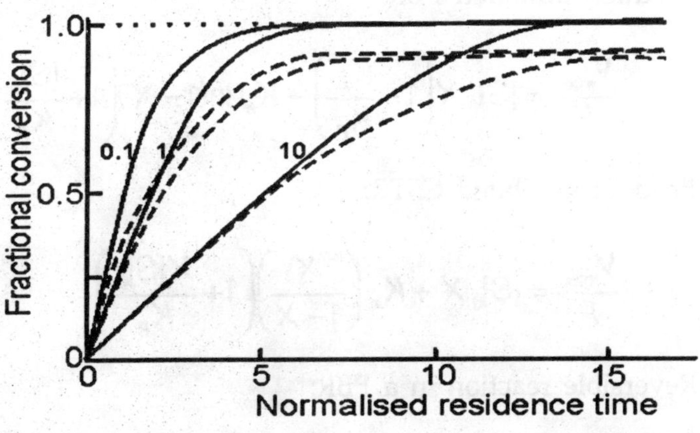

Figure 6. Comparison of the changes in fractional conversion with residence time between the PBR (—) and CSTR (-----------) at different values of $[S]_0/K_m$ (10, 1 and 0.1; higher $[S]_0/K_m$ values giving the lower curves).

The residence time is the reciprocal of the normalised flow rate (Figure 5). If the flow rate is unchanged then the 'normalised residence time' may be thought of as the reactor volume needed to produce the required degree of conversion.

Equations describing the behaviour of CSTRs and PBRs utilising reversible reactions or undergoing product or substrate inhibition can be derived in a similar manner, using equations, and rather than:

Substrate -inhibited PBR:

$$\frac{V_{max}}{F} = [S]_0 X - K_m \ln(1-X) + [S]_0^2 \left(\frac{2X - X^2}{2K_S} \right) \quad (11)$$

Substrate -inhibited CSTR:

$$\frac{V_{max}}{F} = [S]_0 X + K_m \frac{X}{1-X} + [S]_0^2 \left(\frac{X - X^2}{2K_S} \right) \quad (12)$$

Product -inhibited PBR:

$$\frac{V_{max}}{F} = [S]_0 X \left(1 - \frac{K_m}{K_P} \right) - K_m \ln(1-X) \left(1 + \frac{[S]_0}{K_P} \right) \quad (13)$$

Product -inhibited CSTR:

$$\frac{V_{max}}{F} = [S]_0 X + K_m \left(\frac{X}{1-X} \right) \left(1 + \frac{X[S]_0}{K_P} \right) \quad (14)$$

Reversible reaction in a PBR:

$$\frac{V}{F} = X[S^{\#}] - K \ln(1-X) \quad (15)$$

Reversible reaction in a CSTR:

$$\frac{V}{F} = X[S^{\#}] + K\left(\frac{X}{1-X}\right) \tag{16}$$

X in equations (15) and (16) is the fractional conversion for a reversible reaction.

$$X = \frac{[S]_0 - [S]}{[S]_0 - [S]_\infty} \tag{17}$$

The meaning of other symbols used in equations (11)-(17) may be used to compare the size of PBR and CSTR necessary to achieve the same conversion under various conditions (Figure 7). Another useful parameter for comparing these reactors is the productivity. This can be derived for each reactor assuming a first-order inactivation of the enzyme). Combined with equation (6) for PBR, or (10) for CSTR, the following relationships are obtained on integration:

$$\text{PBR: } Ln\left\{\frac{[S]_0 X_0 - K_m Ln(1 - X_0)}{[S]_0 X_t - K_m Ln(1 - X_t)}\right\} = k_d t \tag{18}$$

$$\text{CSTR: } Ln\left\{\frac{[S]_0 X_0 + K_m\left(\dfrac{X_0}{1 - X_0}\right)}{[S]_0 X_t + K_m\left(\dfrac{X_t}{1 - X_t}\right)}\right\} = k_d t \tag{19}$$

where k_d is the first -order inactivation constant (i.e. k_{d1}, in equation (1.25)) and the fractional conversion subscripts refer to time = 0 or t. The change in productivity (Figure 8) and fractional conversion (Figure 9) of these reactors with time can be compared using these equations.

These reactors may be operated for considerably longer periods than that determined by the inactivation of their contained immobilised enzyme, particularly if they are capable of high conversion at low substrate concentrations (Figure 9). This is independent of any enzyme stabilisation and is simply due to such reactors initially containing large amounts of redundant enzyme.

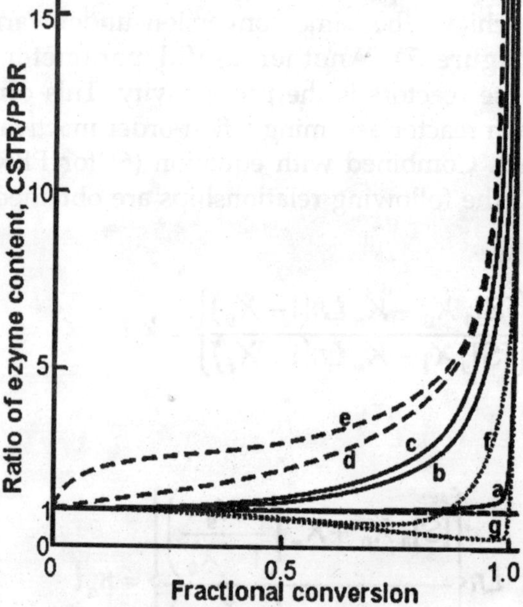

Figure 7. Comparison of the ratio, of the enzyme content in a CSTR to that in a PBR, necessary to achieve various degrees of conversion for a range of process conditions.

The actual size of the CSTR will be five to ten times greater than indicated due to the necessity of maintaining stirring within the vessel. —Uninhibited reaction; ………… product inhibited; ---------substrate inhibited. (curve a) $[S]_0/K_m = 100$; (curve b) $[S]_0/K_m = 1$; (curve c) $[S]_0/K_m < 0.01$; (curve d) $[S]_0/K_m = 1$, product inhibited $K_P/K_m = 0.1$; (curve e) $[S]_0/K_m = 1$, product inhibited $K_P/K_m = 0.01$; (curve f) $[S]_0/K_m = 1$, substrate inhibited $[S]_0/K_S = 10$; (curve g) $[S]_0/K_m = 100$; substrate inhibited $[S]_0/K_S = 10$. The size of a CSTR becomes prohibitively large at high conversions (e.g. using curve b, a CSTR contains three times the enzyme in a PBR to achieve a 90% conversion, but this increases to 18 times for 99% conversion. The difference between the two types of reactor is increased if the effectiveness factor (h) is less than one due to diffusional effects.

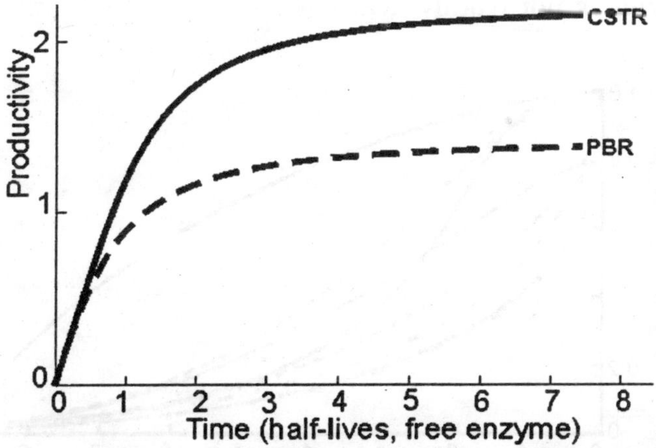

Figure 8. The change in productivity of a PBR (---------) and CSTR (—) with time, assuming an initial fractional conversion (X_0) = 0.99 and $[S]_0/K_m = 100$. The units of time are half -lives of the free enzyme ($t_{1/2} = Ln(2)/k_d$) and the productivity is given in terms of ($FXt_{1/2}$). Although the overall productivity is 1.6 times greater for a CSTR than a PBR, it should be noted that the CSTR contains 1.9 times more enzyme.

In general, there is little or no back -pressure to increased flow rate through the CSTR. Such reactors may be started up as batch reactors until the required degree of conversion is reached, when the process may be made continuous. CSTRs are not generally used in processes involving high conversions but a train of CSTRs may approach the PBR performance. This train may be a number (greater than three) of reactors connected in series or a single vessel divided into compartments, in order to minimise back-mixing CSTRs may be used with soluble rather than immobilised enzyme if an ultrafiltration membrane is used to separate the reactor output stream from the reactor contents. This causes a number of process difficulties, including concentration polarization or inactivation of the enzyme on the membrane but may be preferable in order to achieve a combined reaction and separation process or where a suitable immobilised enzyme is not readily available.

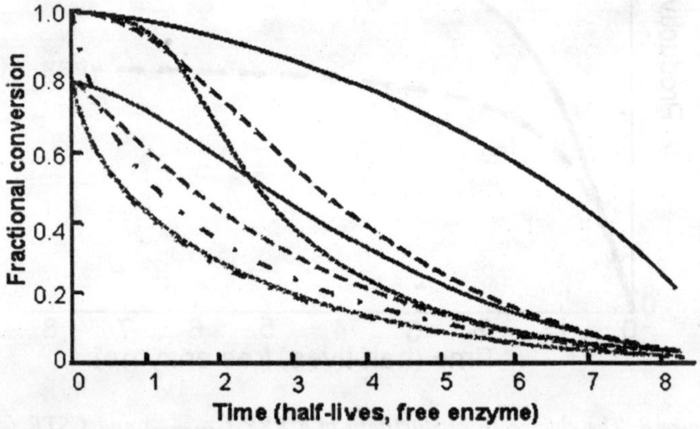

Figure 9. The change in fractional conversion of PBRs and CSTRs with time.

The time is given in terms of the half -life of the free enzyme ($t_{1/2} = Ln(2)/k_d$). Although the CSTR maintains its fractional conversion for a longer period than the PBR, particularly at high X_0. It should be noted that a CSTR capable of $X_0 = 0.99$ at a substrate feed concentration of $[S]_0/K_m = 0.01$ contains 22 times more enzyme than an equivalent PBR, but yields only 2.2 times more product. The initial stability in the fractional conversion over a considerable period of time, due to the enzyme redundancy, should not be confused with any effect due to stabilisation of the immobilised enzyme.

Fluidised Bed Reactors

These reactors generally behave in a manner intermediate between CSTRs and PBRs. They consist of a bed of immobilised enzyme which is fluidised by the rapid upwards flow of the substrate stream alone or in combination with a gas or secondary liquid stream, either of which may be inert or contain material relevant to the reaction. A gas stream is usually preferred as it does not dilute the product stream.

There is a minimum fluidisation velocity needed to achieve bed expansion, which depends upon the size, shape, porosity and density of the particles and the density and viscosity of the liquid. This minimum fluidisation velocity is generally fairly low (about 0.2 -I.0 cm s^{-1}) as most immobilised-enzyme particles have densities close to that of the bulk liquid.

In this case the relative bed expansion is proportional to the superficial gas velocity and inversely proportional to the square root of the reactor diameter. Fluidising the bed requires a large power input but, once fluidised, there is little further energetic input needed to increase the flow rate of the substrate stream through the reactor (Figure 3).

At high flow rates and low reactor diameters almost ideal plug -flow characteristics may be achieved. However, the kinetic performance of the FBR normally lies between that of the PBR and the CSTR, as the small fluid linear velocities allowed by most biocatalytic particles causes a degree of back-mixing that is often substantial, although never total. The actual design of the FBR will determine whether it behaves in a manner that is closer to that of a PBR or CSTR (Figures 5 - 9). It can, for example, be made to behave in a manner very similar to that of a PBR, if it is baffled in such a way that substantial backmixing is avoided. FBRs are chosen when these intermediate characteristics are required, e.g. where a high conversion is needed but the substrate stream is colloidal or the reaction produces a substantial pH change or heat output. They are particularly useful if the reaction involves the utilisation or release of gaseous material.

The FBR is normally used with fairly small immobilised enzyme particles (20-40 mm diameter) in order to achieve a high catalytic surface area. These particles must be sufficiently dense, relative to the substrate stream, that they are not swept out of the reactor. Less-dense particles must be somewhat larger.

For efficient operation the particles should be of nearly uniform size otherwise a non-uniform biocatalytic concentration gradient will be formed up the reactor. FBRs are usually tapered outwards at the exit to allow for a wide range of flow rates. Very high flow rates are avoided as they cause channelling and catalyst loss.

The major disadvantage of development of FBR process is the difficulty in scaling-up these reactors. PBRs allow scale-up factors of greater than 50000 but, because of the markedly different fluidisation characteristics of different sized reactors, FBRs can only be scaled-up by a factor of 10 -100 each time. In addition, changes in the flow rate of the substrate stream causes complex changes in the flow pattern within these reactors that may have consequent unexpected effects upon the conversion rate.

Immobilised-enzyme Processes

Immobilise enzyme systems are used where they offer cost advantages to users on the basis of total manufacturing costs. The plant size needed for continuous processes is two orders of magnitude smaller than that required for batch processes using free enzymes. The capital costs are, therefore, considerably smaller and the plant may be prefabricated cheaply off-site Immobilised enzymes offer greatly increased productivity on an enzyme weight basis and also often provide process advantages. Currently used immobilised-enzyme processes are given in Table 1.

Table 1. Some of the more important industrial uses of immobilised enzymes

Enzyme	EC number	Product
Aminoacylase	3.5.1.14	L-Amino acids
Aspartate ammonia-lyase	4.3.1.1	L-Aspartic acid
Aspartate 4-decarboxylase	4.1.1.12	L-Alanine
Cyanidase (from waste cyanide)	3.5.5.x	Formic acid
Glucoamylase	3.2.1.3	D-Glucose
Glucose isomerase	5.3.1.5	High-fructose corn syrup
Histidine ammonia-lyase	4.3.1.3	Urocanic acid
Hydantoinasea	3.5.2.2	D- and L-amino acids
Invertase	3.2.1.26	Invert sugar
Lactase	3.2.1.23	Lactose-free milk and whey
Lipase	3.1.1.3	Cocoa butter substitutes
Nitrile hydratase	4.2.I.x	Acrylamide

Penicillin amidases	3.5.1.11	Penicillins
Raffinase	3.2.1.22	Raffinose-free solutions
Thermolysin	3.2.24.4	Aspartame

[a] Dihydropyrimidinase.

High-fructose corn syrups (HFCS)

With the development of glucoamylase in the 1940s and 1950s it became a straightforward matter to produce high DE glucose syrups. However, these have shortcomings as objects of commerce: D-glucose has only about 70% of the sweetness of sucrose, on a weight basis, and is comparatively insoluble. Batches of 97 DE glucose syrup at the final commercial concentration (71% (w/w)) must be kept warm to prevent crystallisation or diluted to concentrations that are microbiologically insecure.

Fructose is 30% sweeter than sucrose, on a weight basis, and twice as soluble as glucose at low temperatures so a 50% conversion of glucose to fructose overcomes both problems giving a stable syrup that is as sweet as a sucrose solution of the same concentration. The isomerisation is possible by chemical means but not economical, giving tiny yields and many by-products (e.g. 0.1 M glucose 'isomerised' with 1.22 M KOH at 5°C under nitrogen for 3.5 months gives a 5% yield of fructose but only 7% of the glucose remains unchanged, the majority being converted to various hydroxy acids).

One of the triumphs of enzyme technology so far has been the development of 'glucose isomerase'. Glucose is normally isomerised to fructose during glycolysis but both sugars are phosphorylated. The use of this phosphohexose isomerase may be ruled out as a commercial enzyme because of the cost of the ATP needed to activate the glucose and because two other enzymes (hexokinase and fructose-6-phosphatase) would be needed to complete the conversion.

Only an isomerase that would use underivatised glucose as its substrate would be commercially useful but, until the late 1950s, the existence of such an enzyme was not suspected. At about this time, enzymes were found that catalyse the conversion of D-xylose to an equilibrium mixture of D-xylulose and D-xylose in bacteria. When supplied with cobalt ions, these xylose isomerases were found to isomerise α-D-glucopyranose to α-D - fructofuranose, equilibration from the more abundant β-D-glucopyranose and to the major product β-D-fructopyranose occurring naturally and non-enzymically. Now it is known that several genera of microbes, mainly bacteria, can produce such glucose isomerases:

The commercial enzymes are produced by Actinoplanes missouriensis, Bacillus coagulans and various Streptomyces species; as they have specificities for glucose and fructose which are not much different from that for xylose and ways are being found to avoid the necessity of xylose as inducer, these should perhaps now no longer be considered as xylose isomerases.

They are remarkably amenable enzymes in that they are resistant to thermal denaturation and will act at very high substrate concentrations, which have the additional benefit of substantially stabilising the enzymes at higher operational temperatures. The vast majority of glucose isomerases are retained within the cells that produce them but need not be separated and purified before use.

All glucose isomerases are used in immobilised forms. Although differerent immobilisation methods have been used for enzymes from differerent organisms, the principles of use are very similar. Immobilisation is generally by cross-linking with glutaraldehyde, plus in some cases a protein diluent, after cell lysis or homogenisation.

Originally, immobilised glucose isomerase was used in a batch process. This proved to be costly as the relative reactivity of fructose during the long residence times gave rise to significant by -product production. Also, difficulties were encountered in the removal of the added Mg^{2+} and Co^{2+} and the recovery of the catalyst. Nowadays most isomerisation is performed in PBRs (Table 2). They are used with high substrate concentration (35-45% dry solids, 93-97% glucose) at 55-60°C. The pH is adjusted to 7.5-8.0 using sodium carbonate and magnesium sulphate is added to maintain enzyme activity (Mg^{2+} and Co^{2+} are cofactors). The Ca^{2+} concentration of the glucose feedstock is usually about 25μm, left from previous processing, and this presents a problem. Ca^{2+} competes successfully for the Mg^{2+} binding site on the enzyme, causing inhibition.

At this level the substrate stream is normally made 3 mM with respect to Mg^{2+}. At higher concentrations of calcium a Mg^{2+} : Ca^{2+} ratio of 12 is recommended. Excess Mg^{2+} is uneconomic as it adds to the purification as well as the isomerisation costs. The need for Co^{2+} has not been eliminated altogether, but the immobilisation methods now used fix the cobalt ions so that none needs to be added to the substrate streams.

Table 2. Comparison of glucose isomerisation methods

Parameter	Batch (soluble GI)	Batch (immobilised Gl)	Continuous (PBR)
Reactor volume (m^3)	1100	1100	15
Enzyme consumption (tonnes)	180	11	2
Activity, half-life (h)	30	300	1500
Active life, half-lives	0.7	2	3
Residence time (h)	20	20	0.5
Co2+ (tonnes)	2	1	0
Mg2+ (tonnes)	40	40	7
Temperature (°C)	65	65	60

pH	6.8	6.8	7.6
Colour formation (A_{420})	0.7	0.2	< 0.1
Product refining	Filtration	-	-
	C-treatment[a]	C-treatment	C -
	Cation exchange	Cation exchange	treat-
	Anion exchange	Anion exchange	ment
Capital, labour and energy costs, £ tonne^{-1}	5	5	1
Conversion cost, £ tonne^{-1}	500	30	5

All processes start with 45% (w/w) glucose syrup DE 97 and produce 10000 tonnes per month of 42% fructose dry syrup. Some of the improvement that may be seen for PBR productivity is due to the substantial development of this process.

[a] Treatment with activated carbon.

It is essential for efficient use of immobilised glucose isomerase that the substrate solution is adequately purified so that it is free of insoluble material and other impurities that might inactivate the enzyme by chemical (inhibitory) or physical (pore-blocking) means. In effect, this means that glucose produced by acid hydrolysis cannot be used, as its low quality necessitates extensive and costly purification.

Insoluble material is removed by filtration, sometimes after treatment with flocculants, and soluble materials are removed by ion exchange resins and activated carbon beads. This done, there still remains the possibility of inhibition due to oxidised by-products caused by molecular oxygen. This may be removed by vacuum de-aeration of the substrate at the isomerisation temperature or by the addition of low concentrations (< 50 ppm) of sulphite.

At equilibrium at 60°C about 51 % of the glucose in the reaction mixture is converted to fructose. However, because of the excessive time taken for equilibrium to be attained and the presence of oligosaccharides in the substrate stream, most manufacturers adjust flow rates so as to produce 42-46% (w/w) fructose (leaving 47-51 % (w/w) glucose). To produce 100 tonnes (dry substance) of 42% HFCS per day, an enzyme bed volume of about 4 m^3 is needed. Activity decreases, following a first-order decay equation. The half-life of most enzyme preparations is between 50 and 100 days at 55°C. Typically a batch of enzyme is discarded when the activity has fallen to an eighth of the initial value (i.e. after three half -lives).

To maintain a constant fructose content in the product, the feed flow rate is adjusted according to the enzyme activity. Several reactors containing enzyme preparations of different ages are needed to maintain overall uniform production by the plant (Figure 10). In its lifetime 1 kg of immobilised glucose isomerase will produce 10 -11 tonnes of 42% fructose syrup (dry substance).

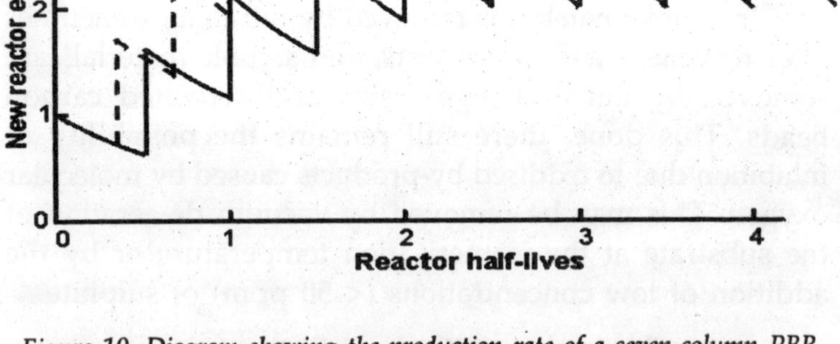

Figure 10. Diagram showing the production rate of a seven-column PBR facility on start-up, assuming exponential decay of reactor activity.

The columns are brought into use one at a time. At any time a maximum of six PBRs are operating in parallel, whilst the seventh, exhausted, reactor is being refilled with fresh biocatalyst. —— PBR activities allowed to decay through three half-lives (to 12.5% initial activity) before replacement. The final average productivity is 2.51 times the initial productivity of one column. - - -- - - - PBR activities allowed to decay through two half-lives (to 25% initial activity) before replacement. The final average productivity is 3.23 times the initial productivity of one column. It may be seen that the final average production rate is higher when the PBRs are individually operated for shorter periods but this 29% increase in productivity is achieved at a cost of 50% more enzyme, due to the more rapid replacement of the biocatalyst in the PBRs. A shorter PBR operating time also results in a briefer start-up period and a more uniform productivity.

After isomerisation, the pH of the syrup is lowered to 4 - 5 and it is purified by ion-exchange chromatography and treatment with activated carbon. Then, it is normally concentrated by evaporation to about 70% dry solids.

For many purposes a 42% fructose syrup is perfectly satisfactory for use but it does not match the exacting criteria of the quality soft drink manufacturers as a replacement for sucrose in acidic soft drinks. For use in the better colas, 55% fructose is required. This is produced by using vast chromatographic columns of zeolites or the calcium salts of cation exchange resins to adsorb and separate the fructose from the other components.

The fractionation process, although basically very simple, is only economic if run continuously. The fructose stream (90% (w/w) fructose, 9% glucose) is blended with 42% fructose syrups to give the 55% fructose (42% glucose) product required. The glucose-rich 'raffinate' stream may be recycled but if this is done undesirable oligosaccharides build up in the system. Immobilised glucoamylase is used in some plants to hydrolyse oligosaccharides in the raffinate; here the substrate concentration is comparatively low (around 20% dry solids) so the formation of isomaltose by the enzyme is insignificant.

Clearly the need for a second large fructose enrichment plant in addition to the glucose isomerase plant is undesirable and attention is being paid to means of producing 55% fructose syrups using only the enzyme. The thermodynamics of the system favour fructose production at higher temperatures and 55% fructose syrups could be produced directly if the enzyme reactors were operated at around 95°C. The use of miscible organic co - solvents may also produce the desired effect. Both these alternatives present a more than considerable challenge to enzyme technology!

The present world market for HFCS is over 5 million tonnes of which about 60% is for 55% fructose syrup with most of the remainder for 42% fructose syrup. This market is still expanding and ensures that HFCS production is the major application for immobilised-enzyme technology.

The high-fructose syrups can be used to replace sucrose where sucrose is used in solution but they are inadequate to replace crystalline sucrose. Another ambition of the corn syrup industry is to produce sucrose from starch. This can be done using a combination of the enzymes phosphorylase (EC 2.4.1.1), glucose isomerase and sucrose phosphorylase (EC 2.4.1.7), but the thermodynamics do not favour the conversion so means must be found of removing sucrose from the system as soon as it is formed. This will not be easy but is achievable if the commercial pull (i.e. money available) is sufficient:

$$\text{starch } (G_n) + \text{orthophosphate} \xrightleftharpoons{\text{phosphorylase}} \text{starch } (G_{n-1}) + \alpha\text{-glucose-1-phosphate}$$

glucose isomerase

glucose ⇌ fructose sucrose phosphorylase

a-glucose-1 -phosphate + fructose ⇌ sucrose + orthophosphate

A further possible approach to producing sucrose from glucose is to supply glucose at high concentrations to microbes whose response to osmotic stress is to accumulate sucrose intracellularly. Provided they are able to release sucrose without hydrolysis when the stress is released, such microbes may be the basis of totally novel processes.

Immobilised Raffinase

The development of a raffinase (α-D-galactosidase) suitable for commercial use is another triumph of enzyme technology. Plainly, it would be totally unacceptable to use an enzyme preparation containing invertase to remove this material during sucrose production. It has been necessary to find an organism capable of producing an a-galactosidase but not an invertase.

A mould, Mortierella vinacea var. raffinoseutilizer, fills the requirements. This is grown in a particulate form and the particles harvested, dried and used directly as the immobilised-enzyme preparation. It is stirred with the sugar beet juice in batch stirred tank reactors. When the removal of raffinose is complete, stirring is stopped and the juice pumped off the settled bed of enzyme.

Enzyme, lost by physical attrition, is replaced by new enzyme added with the next batch of juice. The galactose released is destroyed in the alkaline conditions of the first stages of juice purification and does not cause any further problems while the sucrose is recovered. This process results in a 3% increase in productivity and a significant reduction in the costs of the disposal of waste molasses.

Immobilised raffinase may also be used to remove the raffinose and stachyose from soybean milk. These sugars are responsible for the flatulence that may be caused when soybean milk is used as a milk substitute in special diets.

Immobilised Invertase

Invertase was probably the first enzyme to be used on a large scale in an immobilised form. In the period 1941 - 1946 the acid, previously used in the manufacture of Golden Syrup, was unavailable, so yeast invertase was used instead. Yeast cells were autolysed and the autolysate clarified by adjustment to pH 4.7, followed by filtration through a bed of calcium sulphate and adsorption into bone char. A layer of the bone char containing invertase was included in the bed of bone char already used for decolourising the syrup.

The scale used was large, the bed of invertase-char being 2 ft (60 cm) deep in a bed of char 20 ft (610 cm) deep. The preparation was very stable, the limiting factors being microbial contamination or loss of decolourising power rather than loss of enzymic activity. The process was cost-effective but, not surprisingly, the product did not have the subtlety of flavour of the acid-hydrolysed material and the immobilised enzyme process was abandoned when the acid became available once again.

Recently, however, it has been relaunched using Brimac™, where the invertase -char mix is stabilised by cross-linking and has a half-life of 90 days in use (pH 5.5, 50°C). The revival is due, in part, to the success of HFCS as a high-quality low-colour sweetener. It is impossible to produce inverted syrups of equivalent quality by acid hydrolysis.

Enzymic inversion avoids the high-colour, high salt-ash, relatively low conversion and batch variability problems of acid hydrolysis. Although free invertase may be used (with residence times of about a day), the use of immobilised enzymes in a PBR (with residence time of about 15 min) makes the process competitive; the cost of 95% inversion (at 50% (w/w)) being no more than the final evaporation costs (to 75% (w/w)). A productivity of 16 tonnes of inverted syrup (dry weight) may be achieved using one litre of the granular enzyme.

Production of Amino Acids

Another early application of an immobilised enzyme was the use of the aminoacylase from Aspergillus oryzae to resolve racemic mixtures of amino acids.

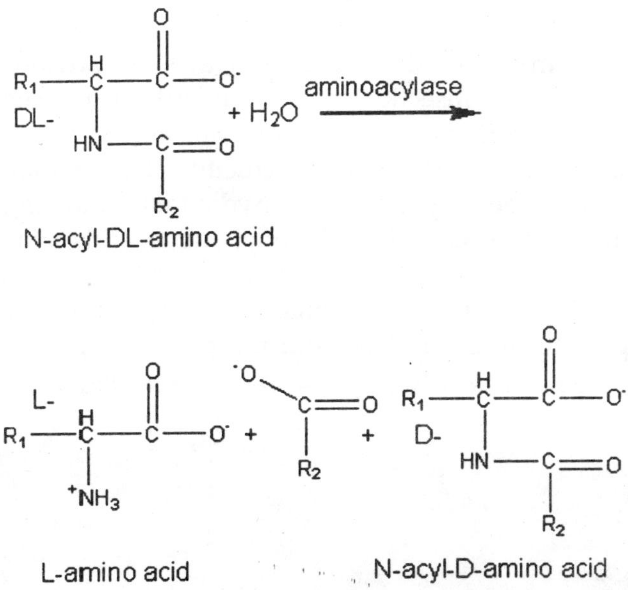

Chemically synthesised racemic N-acyl-DL-amino acids are hydrolysed at pH 8.5 to give the free L-amino acids plus the unhydrolysed N-acyl-D-amino acids. These products are easily separated by differential crystallisation and the

N-acyl-D-amino acids racemised chemically (or enzymically) and reprocessed. The enzyme is immobilised by adsorption to anion exchange resins (e.g. DEAE-Sephadex) and has an operational half-life of about 65 days at 50°C in PBRs with residence times of about 30 min. The reactors may be re-activated in situ by simply adding more enzyme. The immobilised enzyme has proved a more economical process than the use of free enzyme mainly due to the more efficient use of the substrate and reductions in the cost of enzyme and labour.

Novel and natural L-amino acids can be produced by the chemical conversion of aldehydes through DL-amino nitrites to racemic DL-hydantoins followed by enzymic hydrolysis with hydantoinase and a carbamoylase at pH 8.5. Both enzymes may be obtained from Arthrobacter species.

D-Amino acids are important constituents in antibiotics and insecticides. They may be produced in a manner similar to the L-amino acids but using hydantoinases of differing specificity. The Pseudomonas striata enzyme is specific for D-hydantoins, allowing their specific hydrolysis to D-carbamoyl amino acids which can be converted to the D-amino acids by chemical treatment with nitrous acid. They remaining L-hydantoin may be simply racemised by base and the process repeated.

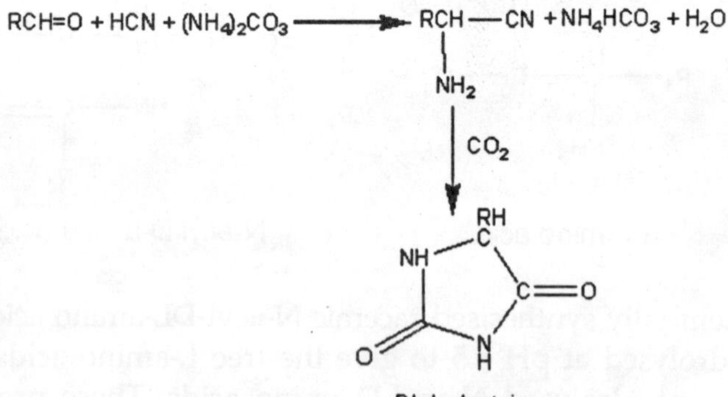

DL-hydantoin

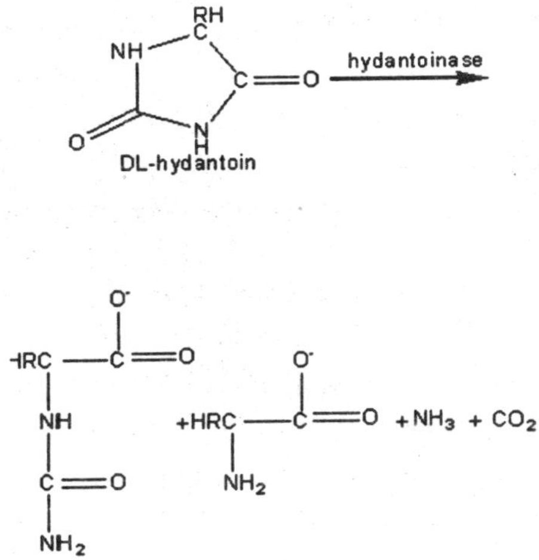

L-Aspartic acid is widely used in the food and pharmaceutical industries and is needed for the production of the low -calorific sweetener aspartame. It may be produced from fumaric acid by the use of the aspartate ammonia-lyase (aspartase) from Escherichia coli.

aspartate ammonia-lyase

$$^{-}OOCCH{=}CHCOO^{-} {}_{+} NH_4{+} \rightleftharpoons {}^{-}OOCCH_2\,CH(NH_3{+})COO^{-}$$

fumaric acid L-aspartic acid

A crude immobilised aspartate ammonia-lyase (50000 Ug^{-1}) may be prepared by entrapping Escherichia coli cells in a κ-carageenan gel crosslinked with glutaraldehyde and hexamethylenediamine. The process is operated in a PBR at pH 8.5 using ammonium fumarate as the substrate, with a reported operational half-life of 680 days at 37°C.

Urocanic acid is a sun-screening agent which may be produced from L-histidine by the histidine ammonia -lyase (histidase) from Achromobacter liquidum. The organism cannot be used directly as it has urocanate hydratase activity, which removes the urocanic acid. However, a brief heat treatment (70°C, 30 min) inactivates this unwanted activity but has little effect on the histidine ammonia-lyase. A crude immobilised-enzyme preparation consisting of heat -treated cells entrapped in a polyacrylamide gel has been used to effect this conversion, showing a half-life of 180 days at 37°C.

Immobilised Lactase

Lactase is one of relatively few enzymes that have been used both free and immobilised in large-scale processes. The reasons for its utility has been given earlier, but the relatively high cost of the enzyme is an added incentive for its use in an immobilised state.

Immobilised lactases are important mainly in the treatment of whey, as the fats and proteins in the milk emulsion tend to coat the biocatalysts. This both reduces their apparent activity and increases the probability of microbial colonisation.

Yeast lactase has been immobilised by incorporation into cellulose triacetate fibres during wet spinning, a process developed by Snamprogetti S.p.A. in Italy. The fibres are cut up and used in a batchwise STR process at 5°C (Kluyveromyces lactis, pH optimum 6.4 -6.8, 90 Ug^{-1}). Fungal lactases have been immobilised on 0.5 mm diameter porous silica (35 nm mean pore diameter) using glutaraldehyde and γ-aminopropyltriethoxysilane (Asperigillus niger, pH optimum 3.0 -3.5, 500 U g-1; A. oryzae, pH optimum 4.0 -1.5, 400 Ug^{-1}). They are used in PBRs. Due to the different pH optima of fungal and yeast lactases, the yeast enzymes are useful at the neutral pH of both milk and sweet whey, whereas fungal enzymes are more useful with acid whey.

Immobilised lactases are particularly affected by two inherent short-comings. Product inhibition by galactose and unwanted oligosaccharide formation are both noticeable under the diffusion-controlled conditions usually prevalent.

Both problems may be reduced by an increase in the effectiveness factor and a reduction in the degree of hydrolysis or initial lactose concentration, but such conditions also lead to a reduction in the economic return. The control of microbial contamination within the bioreactors is the most critical practical problem in these processes. To some extent, this may be overcome by the use of regular sanitation with basic detergent and a dilute protease solution.

Production of Antibiotics

Benzylpenicillins and phenoxymethylpenicillins (penicillins G and V, respectively) are produced by fermentation and are the basic precursors of a wide range of semi-synthetic antibiotics, e.g. ampicillin. The amide link may be hydrolysed conventionally but the conditions necessary for its specific hydrolysis, whilst causing no hydrolysis of the intrinsically more labile but pharmacologically essential b-lactam ring, are difficult to attain. Such specific hydrolysis may be simply achieved by use of penicillin amidases (also called penicillin acylases). Different enzyme preparations are generally used for the hydrolysis of the penicillins G and V, pencillin-V-amidase being much more specific than pencillin-G-amidase.

Penicillin amidase may be obtained from E. coli and has been immobilised on a number of supports including cyanogen bromide-activated Sephadex G200. It represents one of the earliest successful processes involving immobilised enzymes and is generally used in batch or semicontinuous STR processes (40,000 Ukg^{-1} penicillin G,

35°C, pH 7.8, 2 h) where it may be reused over 100 times. It has also been used in PBRs, where it has an active life of over 100 days, producing about two tonnes of 6-aminopenicillanic acid kg-1of immobolised enzyme.

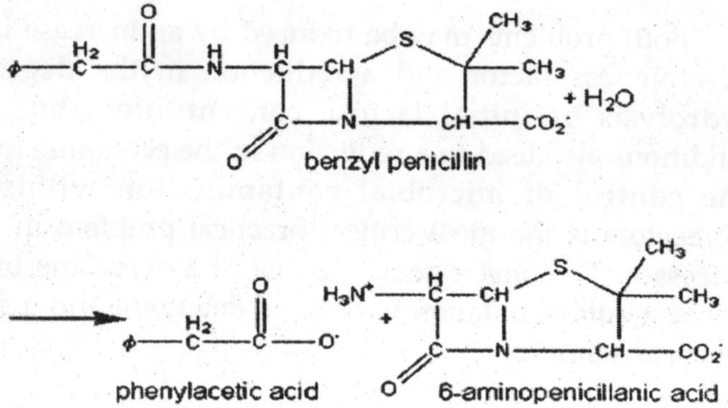

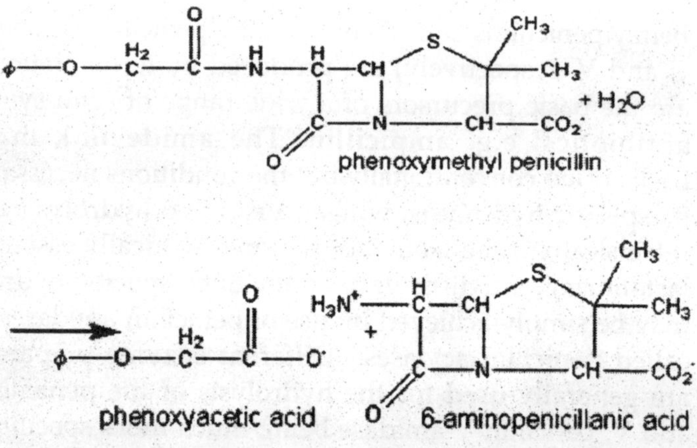

The penicillin-G-amidases may be used 'in reverse' to synthesise penicillin and cephalosporin antibiotics by non -equilibrium kinetically controlled reactions. Ampicillin has been produced by the use of penicillin-G-amidase immobilised by adsorption to DEAE -cellulose in a packed bed column:

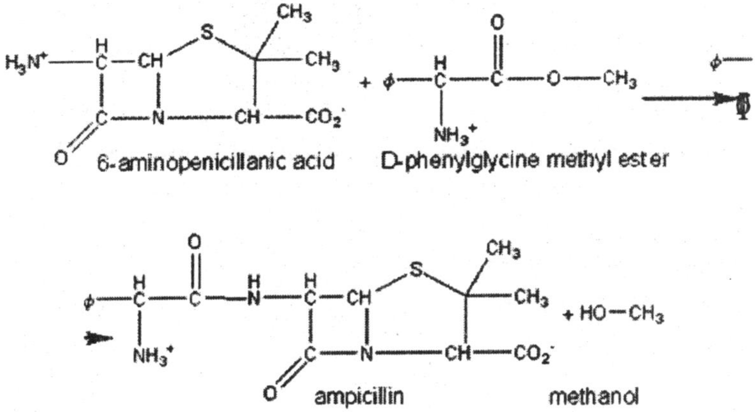

Many other potential and proven antibiotics have been synthesised in this manner, using a variety of synthetic b-lactams and activated carboxylic acids.

Preparation of Acrylamide

Acrylamide is an important monomer needed for the production of a range of economically useful polymeric materials. It may be produced by the addition of water to acrylonitrile.

$$CH_2{=}CHCN + H_2O \longrightarrow CH_2{=}CHCONH_2$$

This process may be achieved by the use of a reduced copper catalyst (Cu^+); however, the yield is poor, unwanted polymerisation or conversion to acrylic acid ($CH_2{=}CHCOOH$) may occur at the relatively high temperatures involved (80 -140°C) and the catalyst is difficult to regenerate. These problems may be overcome by the use of immobilised nitrile hydratase (often erroneously called a nitrilase). The enzyme from Rhodococcus has been used by the Nitto Chemical

Industry Co. Ltd, as it contains only very low amidase activity which otherwise would produce unwanted acrylic acid from the acrylamide.

Immobilised nitrile hydratase is simply prepared by entrapping the intact cells in a cross-linked 10% (w/v) polyacrylamide/dimethylaminoethylmethacrylate gel and granulating the product. It is used at 10°C and pH 8.0-8.5 in a semibatchwise process, keeping the substrate acrylonitrile concentration below 3% (w/v). Using 1% (w/v) immobilised-enzyme concentration (about 50,000 U l-1) the process takes about a day. Product concentrations of up to 20% (w/v) acrylamide have been achieved, containing negligible substrate and less than 0.02% (w/w) acrylic acid. Acrylamide production using this method is about 4000 tonnes per year.

The closely related enzymes cyanidase and cyanide hydratase are used to remove cyanide from industrial waste and in the detoxification of feeds and foodstuffs containing amygdalin.

$$HCN + 2H_2O \longrightarrow HCOO^- + NH_4^+$$

$$HCN + H^2O \longrightarrow HCONH_2$$

Cell and Enzyme Immobilisation

Enzymes are protein catalysts. After synthesis within a cell, enzymes can function independently of the cell provided certain conditions are maintained. Enzyme technology involves the production, isolation and purification of enzymes. Commercial enzymes are obtained generally from

(a) Plants

(b) Animals

(c) Microbes

Most enzymes are obtained from microbes (fungi and bacteria). Mutant strains have been selected which maximise enzyme production. Genetic engineering has come to the fore in the tailoring of producer organisms for synthesis. Over 2000 enzymes have been isolated to date, yet only a small number are produced on a large scale. Relatively crude preparations used in the food processing industries and in the manufacture of detergents make up the bulk of the output.

Many commercially produced enzymes are hydrolases such as amylases, cellulases, pectinases, and peptidases. Nevertheless, there is a growing demand for purified enzymes such as glucose oxidase for the biochemical analysis of blood glucose concentration and asparaginase for the treatment of a form of leukaemia.

Procedure for Enzyme Production:

1. Selected strains of microorganisms are taken for the culture.
2. They are inoculated into large sterilised vessels containing previously sterilised and cooled nutrient medium.
3. Sterile air is pumped through and the bioreactor is cooled with circulating water. The optimum temperature is between 18°C and 37°C.

Extraction of Enzymes

1. The nutrient liquor is separated into:
   (i) liquid with the dissolved contents and
   (ii) the insoluble cells and other substances.

If the enzyme is intracellular the cells are separated from the contents of the growth medium by centrifuge or filtration. The cells are then disrupted by ultrasound or liquidising and the contents separated from the membranes/walls again by filtration. If the enzyme is extracellular separation from the liquid is by centrifuge or filtration.

The enzyme solution is now concentrated by removing most of the water by one of the following methods:

— low temperature vacuum evaporation

— reverse osmosis.

3. Preservatives are then added and the enzyme is then packaged for distribution to industrial users.
4. If pure enzymes are needed then other procedures are involved before packaging, e.g. fractional precipitation.

Enzyme Immobilisation

The widespread use of enzymes brought with it a problem - how to recover the enzyme for reuse. Many enzymes are expensive to produce and extract. The answer came with recent development of cell / enzyme immobilisation. Immobilisation is the conversion of: Enzyme (water soluble)

Enzyme (water insoluble)

The enzyme still however maintains its catalytic activity. There are four ways of bringing about enzyme immobilisation.

1. Adsorption onto insoluble surfaces e.g. charcoal, clay (Fig. 11).

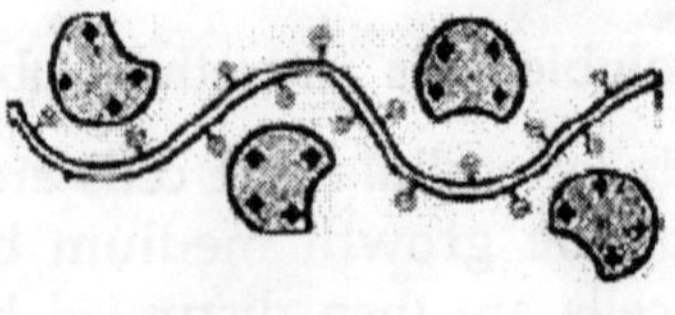

Figure 11.

2. Covalent attachment to collagen (Fig. 12). Not the easiest method.

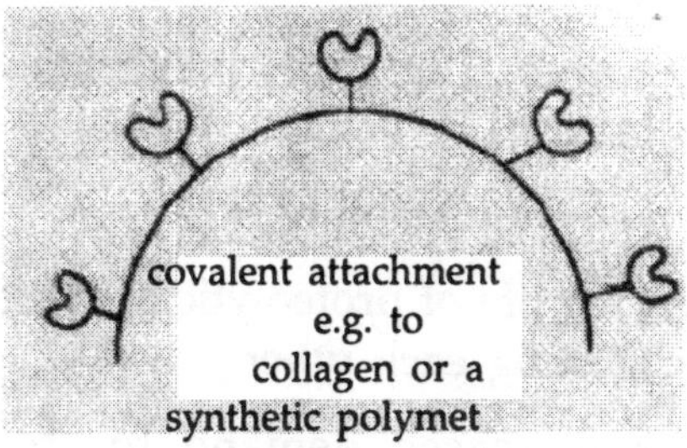

Figure 12.

3. Entrap the enzymes in gels (Fig. 13). The enzymes are trapped in the fibres of the gels. The disadvantage of this method is that substrate or product may not be small enough to enter or leave the gel.

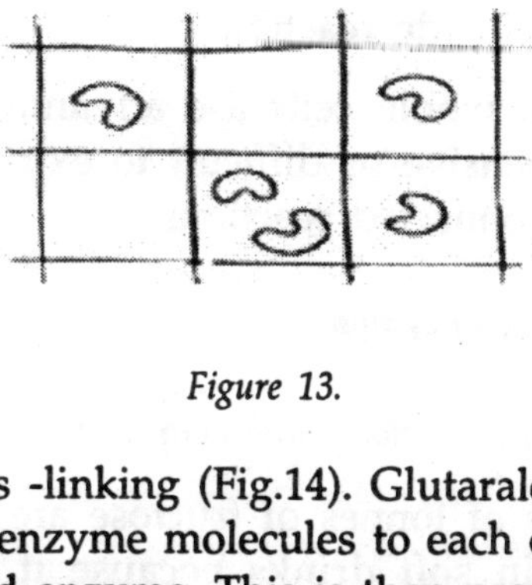

Figure 13.

4. Direct cross -linking (Fig.14). Glutaraldehyde is used to join the enzyme molecules to each other to form a polymerised enzyme. This is then precipitated out of solution and is thus immobilised without the need for a solid support.

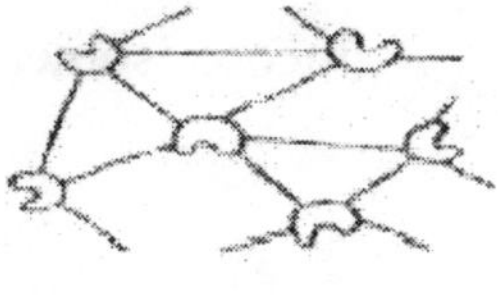

Figure 14.

Advantages of Immobilized Enzymes

1. Easy recovery of enzymes for reuse.
2. Easy harvesting of products (no enzyme contamination).
3. Greater enzyme stability.
4. Extends the life of proteolytic enzymes by preventing them digesting each other.

There is an increased demand for cell-free enzymes because:

— When whole cells are used some of the substrate is used in the cell metabolism

— The optimum condition for individual enzymes is often different to that of a whole cell.

— No wasteful side reactions.

In some cases, whole cells are advantageous when the enzyme is expensive or difficult to extract or involves a chain of interconnected reactions.

Uses of Industrial Enzymes

1. The production of fructose from corn for use in canned drinks.

Many millions of tonnes of fructose are produced every year for use in soft drinks because it is sweeter than glucose or sucrose.

amylase, amyloglucosidase

Corn Starch   Glucose

glucose isomerase

Glucose   Fructose

Glucose isomerase is not an easy enzyme to produce therefore it is economical to be able to reuse this enzyme again

1. The crude enzyme is mixed with gelatine and heated to 40°C.
2. The mixture is sprayed onto a cold immiscible solvent e.g. butyl acetate.
3. The gelatine with the enzyme solidifies and falls to the bottom of the column.
4. The particles are dehydrated in alcohol.
5. The particles are cross-linked with glutaraldehyde.
6. The particles are then washed and put into a large column.
7. Glucose is then poured into the top of the column; it is converted to fructose, which is harvested at the bottom free from contamination of the enzyme. The enzyme remains active for 100 days.
8. Its activity is monitored regularly by testing for the presence of glucose.

2. To make lactose-free milk

A large percentage of the human population are lactose intolerant. They are missing an enzyme, lactase, which converts lactose to glucose and galactose. When lactose is consumed in the diet it remains undigested and passes into the colon where bacteria may convert it into a variety of different chemicals e.g. ethanoic acid, butanoic acid, carbon dioxide, and methane. These cause a variety of symptoms - cramps, distension, acid diarrhoea. Lactose is removed from milk to make it acceptable to those who are lactose intolerant.

Heat-treated skimmed milk is passed through a column containing immobilised lactase. The lactose- free milk emerges at the bottom and is then packaged for consumers.

3. Clarification of fruit juices

Fruit juices contain pectins, which are a complex group of carbohydrates that cement adjacent cells together. They occur in the middle lamella and in the presence of calcium ions form gels. This poses a viscosity problem for fruit juice manufacturers. Also, pectins can form colloids with proteins giving an unwelcome haze in the juice, for example in apple juice.

Immobilised pectinase is used to remove pectin from the juices.

4. Meat tenderisation

In olden days freshly slaughtered meat was hung for several days or weeks to allow the meat to tenderise. Proteolytic enzymes released from lysosomes gradually broke down the muscle structure.

Today, immobilised enzymes are used to shorten the hanging time. Many fungi, bacteria, and plants e.g. pineapple, contain proteolytic enzymes, which can be used for this purpose. These are non-toxic and are capable of tenderising the meat. However the difficulty with the use of these immobilised enzymes is that, if the enzyme is injected into the meat after slaughter, it causes uneven tenderisation.

The most effective way of introducing these enzymes to the muscle tissue is to inject the enzyme into the animal's jugular vein before slaughter (usually up to half an hour prior to slaughter).

5. Production of vinegar

Ethanoic acid is produced by the following reaction:

$$C_2H_5OH + O_2\ CH_3COOH + H_2O.$$

In the fermenter, the bacteria are immobilised on wood shavings or some other inert material while the ethanol is sprinkled over them. Air is forced upwards through the shavings and vinegar is drawn off at the bottom.

Diagnostic Reagents

Enzymes are immobilised onto reagent dip strips for diagnostic processes e.g. in glucose concentration determination in blood samples. The enzymes glucose oxidase and peroxidase are immobilised.

10

Advances in Enzyme Technology

Enzymic Reactions in Biphasic Liquid Systems

It would often be useful if enzyme catalysed reactions could be performed in solvents other than water, as this is not the ideal medium for the majority of organic reactions. Many reactants (e.g. molecular oxygen, steroids and lipids) are more soluble in organic solvents than in water and some products may be quite labile in an aqueous environment.

Accomplishing reactions with such substrates (e.g. the aerobic oxidation of oestrogens catalysed by fungal laccase) in non-aqueous media allows a much increased volumetric activity to be achieved. Microbial contamination, by contrast, is much less of a problem in such solvents, and the consequent absence of microbial proteases may lead to an apparent stabilisation in the biocatalyst.

Some polymerising reactions, for example the polymerisation of phenols catalysed by peroxidase, will produce a higher molecular weight product when carried out in a solution more able to dissolve the products (i.e. oligomers) initially formed. Under normal physiological conditions, hydrolytic enzymes catalyse the degradation of polymers; i.e. hydrolases are transferases normally transferring a moiety to the acceptor, water.

Water is normally present in a vast molar excess over other potential acceptor molecules so no reaction occurs other than hydrolysis. Also, the normal 'concentration' of water (about 55.5 M) is much greater than its typical Km (about 50 mM) and the rate of hydrolysis will not be affected as the reaction proceeds.

By greatly reducing the water activity in these systems they can be used to transfer to other acceptors. Examples of this can be found in the transesterification reactions of esterases and lipases, described more fully later, and the (undesirable) formation of isomaltose from glucose catalysed by glucoamylase.

Restriction of the enzyme to the aqueous phase effectively immobilises the enzyme and allows its straightforward separation, using phase separators developed for the established chemical process industry, from product-containing organic phase. The main asset of these systems, however, is their ability to shift the thermodynamic equilibria of the reactions. Enzymes do not change the equilibrium constants (K_{eq}) of reactions.

Although changes in the physical conditions (i.e. temperature, pH and pressure) do affect the K_{eq} of a reaction, usually this effect is relatively slight over the physical range allowed by stability of the biocatalysts. Use of a biphasic aqueous-organic system, however, may result in substantial changes in the practically useful 'apparent' K_{eq}.

The use of enzymes within organic solvents normally results in a two phase system as all water-soluble enzymes possess a significant amount of strongly bound water, even when in an apparently dry state. This is shown schematically in Figure 1.

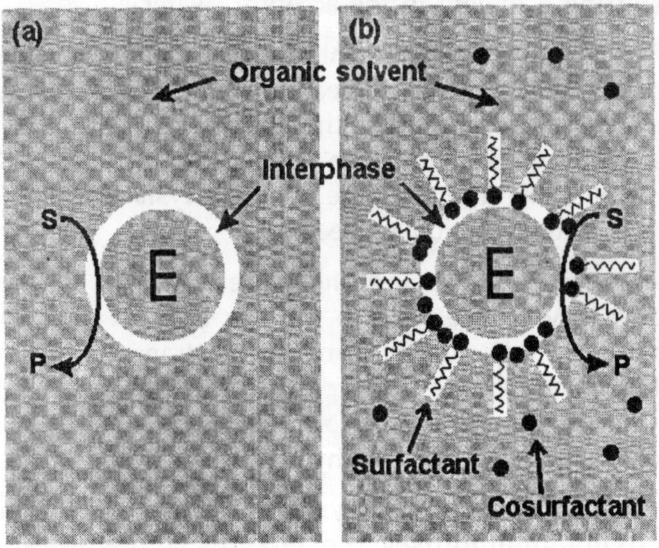

Figure 1. Schematic diagram showing two configurations for an enzyme within an organic solvent.

In the above figure (a) almost-anhydrous enzyme suspended in the organic solvent. The enzyme (E) is surrounded by a thin interphase consisting of water or water plus immobilisation support. (b) enzyme dissolved in a reversed micellar medium. The micelles are formed by the surfactant molecules with assistance from the cosurfactant. The surfactant (e.g. cetyltrimethylammonium bromide (CTAB), bis(2-ethylhexyl) sodium sulphosuccinate (AOT), phosphatidylcholine, tetra-ethylene-glycoldodecylether) is only found at the interphase boundary, whereas the water-immiscible cosurfactant (e.g. butanol, hexanol, octanol), added to vary the properties of this interphase, is generally less polar and more soluble in the organic continuous phase. Both preparations (a) and (b) give optically transparent solutions.

Stabilisation of Enzymes in Biphasic Aqueous-organic Systems

It should become clear from the later discussion that there may be a substantial advantage to be gained from the use of biphasic systems in many enzyme-catalysed reactions. One major factor must first be addressed; the stability of the enzyme in these systems. A distinction should be drawn between the more water-soluble hydrophilic enzymes and the more hydrophobic enzymes often associated with lipid and membranes (e.g. lipases).

The active integrity and stability of hydrophilic enzymes appears to depend on the presence of a thin layer of water, just a few molecules thick, within the microenvironment. This amount of water is miniscule (between 50 and 500 molecules of water for each enzyme molecule) and the enzyme may effectively be operating in an almost anhydrous state. Some hydrophobic lipases retain activity even if fewer molecules of water remain; presumably just sufficient to stabilise the conformation of the active site. The pH of such minute pools of water, containing no free hydrogen ions, is impossible to measure, or control, directly.

However, it appears that the enzyme 'remembers' the pH of its last aqueous solution and functions as though at that pH. If the enzyme-bound water is stripped out or diluted by the use of the more water-soluble, or miscible, organic solvents then the enzyme is usually inactivated. However, under conditions where this does not occur, the limited amount of water available, and the associated reduction in the water activity, considerably reduces the rate of thermoinactivation. This has a stabilising effect on most enzymes; porcine pancreatic lipase, for example, has a half-life of greater than 12 hours at 100°C in 0.02% water in tributyrin, whereas this drops to 12 minutes at a 0.8%

water content and inactivation is almost instantaneous in 100% water.

Additionally, the freezing point of the water is reduced which allows the use of particularly heat-labile enzymes at very low temperatures. The lowering of the water activity tends to produce a more rigid enzyme molecule which may affect both its Km and Vmax. In extreme cases, this may result in a change in the catalytic properties. Porcine pancreatic lipase demonstrates this effect. When used in biphasic systems of low water activity, it no longer catalyses transesterification reactions involving the bulky tertiary alcohols.

The most important factor in the balance between stabilisation and inactivation, due to organic phase, is the solvent polarity. Solvents of lower polarity (i.e. greater hydrophobicity) are less able to disrupt the structure of the necessary tightly bound water molecules. The best measure of polarity is the logarithm of the partition coefficient (LogP) of the organic liquid between n-octanol and water; the higher the LogP, the more non-polar (hydrophobic) is the solvent (Table 1).

$$\text{Log}P = \text{Log}_{10}\left(\frac{[Material_{octanol}]}{[Material_{water}]}\right)$$

Table 1. LogP values of the more commonly used organic solvents

Solvent	LogP	Solvent	LogP
Butanone	0.3	1,1,1-trichloroethane	2.8
Ethyl acetate	0.7	Carbon tetrachloride	2.8
Butanol	0.8	Dibutyl ether	2.9
Diethyl ether	0.8	Cyclohexane	3.1
Methylene chloride	1.4	Hexane	3.5
Butyl acetate	1.7	Petroleum ether (60-80)	3.5

Di-isopropyl ether	2.0	Petroleum ether (80-100)	3.8
Benzene	2.0	Dipentyl ether	3.9
Chloroform	2.2	Heptane	4.0
Tetrachloroethylene	2.3	Petroleum ether (100-120)	4.3
Toluene	2.7	Hexadecane	8.7

LogP values increase by about 0.52 for every methylene group ($-CH_2-$) added in an homologous series. Thus, the LogP of hexanol is that of butanol (0.8) plus 2 x 0.52 (i.e. approximately 1.8).

There appears to be a clear correlation between the activity of biocatalysts in two-phase systems and the LogP (Figure 2). The S-shape of this relationship suggests that enzymes are generally inactivated by solvents with logP < 2 but are little affected by solvents with LogP > 4. There is some variation in the effects between different enzymes and different solvents which makes activity prediction, in the LogP range of 2 - 4, rather imprecise. This range includes some of the most utilised organic phases (e.g. chloroform), which may be suitable for some applications but cause harmful inactivation in others.

The solubility of the reactant(s) and product(s) may considerably reduce the range of LogP that are available for a particular application; many basically nonpolar molecules possessing some polar structural regions which cause their lack of solubility in strongly hydrophobic solvents. The choice of organic phase will also depend on additional factors such as cost, ease of recovery, fire and fume hazards and specific inhibitory effects.

The S-shaped curve can be shifted to the left by immobilising the enzyme within a highly hydrophilic support (Figure 2). A simple way of achieving this is to impregnate a beaded hydrophilic polymer (e.g. Sephadex, agarose) with the enzyme followed by suspension of the wet beads directly in the organic phase. These shifts are very important as they greatly increase the choice of suitable organic phase. Such impregnated beads have the additional advantages that

a. they protect the contained enzyme from liquid-liquid interfacial denaturation at higher rates of stirring;

b. they enable more facile recovery of the biocatalyst;

c. they may be used with low molecular weight hydrophilic coenzymes (e.g. $NAD(P)^+$) with the assistance of a coenzyme regenerating process; the coenzyme being effectively immobilised within the aqueous pools; and

d. they allow the efficient and continuous use of biphasic PBRs, so long as the moving organic phase remains saturated with water.

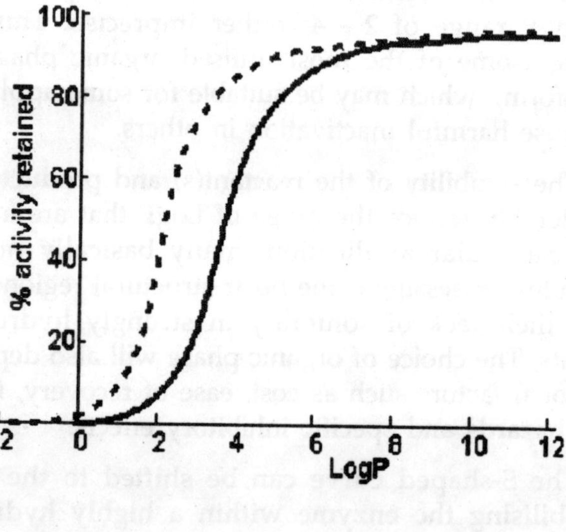

Figure 2. Schematic diagram showing the dependence of the activity of immobilised enzymes, in biphasic systems, on the LogP of the organic phase. --- free enzyme; -------- enzyme immobilised within a strongly hydrophilic support.

Biphasic systems may be further stabilised by the use of deuterated water (D_2O). This reduces the rate of thermal inactivation, although it does cause an increase in the pK_as of ionising groups by about 0.4 with the associated changes in the pH-activity and pH-stability relationships. The higher cost of the deuterated water is offset, to a certain extent, by the small amount necessary in these systems and the ease with which it may be recovered at the end of a catalytic process.

Even where the organic solvent has very low LogP and is miscible, the effect of the expected loss in enzymic activity may be offset by changes in the equilibrium constant. Thus it has been proposed that glucose isomerase be used in aqueous ethanol to produce high fructose corn syrup. The equilibrium fraction of fructose can be raised by about 10% to 55% at 30°C in 80% v/v ethanol. This is economically valid even though the enzymic activity drops by about 10% compared to that in the absence of ethanol.

Equilibria

Much of the theory outlining the effect of biphasic systems on the apparent equilibria was outlined by Karel Martinek and his co-workers at Lomonosov Moscow State University. Consider the simple reaction scheme, involving the equilibrium between reactant (A) and product (B) with K_w representing the equilibrium constant (K_{eq}) in water:

$$A \underset{}{\overset{K_w}{\rightleftharpoons}} B \qquad (1)$$

If this reaction is carried out in a biphasic system consisting of a mixture of an aqueous and an organic solvent, both A and B will be partitioned between the two solvents (with partition coefficients, P_A and P_B) and a separate equilibrium (K_{org}) will be established between A and B in the organic phase.

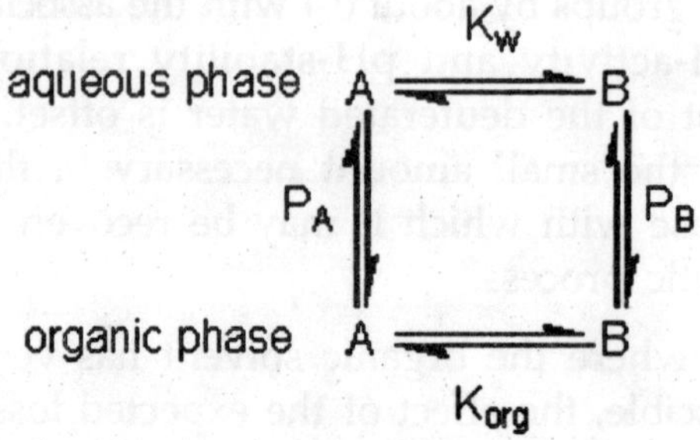

(2)

where:

$$K_W = \frac{[B_W]}{[A_W]} \qquad (2)$$

$$K_{org} = \frac{[B_{org}]}{[A_{org}]} \qquad (3)$$

$$P_A = \frac{[A_{org}]}{[A_w]} \qquad (4)$$

$$P_B = \frac{[B_{org}]}{[B_w]} \qquad (5)$$

Because of the cyclic nature of these equilibria only three of these parameters are independent 'variables' (i.e. K_{org} depends on K_w, P_A and P_B). It should be noticed from the following discussion that K_{org} plays no further part, although similar derivations could be made involving K_{org}, P_A and P_B rather than K_w, P_A and P_B. The apparent equilibrium constant ($K_{biphasic}$) of this system may be defined as

$$K_{biphasic} = \frac{[B_t]}{[A_t]} \tag{6}$$

where the subscripts t, org and w refer to the total solution, the organic and water phases respectively. If V represents the volumes involved, it follows that

$$[A_t]V_t = [A_w]V_w + [A_{org}]V_{org} \tag{7}$$

$$[B_t]V_t = [B_w]V_w + [B_{org}]V_{org} \tag{8}$$

$$V_t = V_w + V_{org} \tag{9}$$

Substituting from equations 7 and 8 into equation 6,

$$K_{biphasic} = \frac{[B_t]V_t}{[A_t]V_t} = \frac{[B_w]V_w + [B_{org}]V_{org}}{[A_w]V_w + [A_{org}]V_{org}} \tag{10}$$

$$K_{biphasic} = \frac{\frac{[B_w]V_w}{[A_w]V_w} + \frac{[B_{org}]V_{org}}{[A_w]V_w}}{\frac{[A_w]V_w}{[A_w]V_w} + \frac{[A_{org}]V_{org}}{[A_w]V_w}} = \frac{\frac{[B_w]}{[A_w]} + \frac{[B_{org}]}{[B_w]}\frac{[B_w]}{[A_w]}\frac{V_{org}}{V_w}}{1 + \frac{[A_{org}]}{[A_w]}\frac{V_{org}}{V_w}} \tag{11}$$

Substituting the partition coefficients from equations 3 and 4 gives the simplified relationship,

$$K_{biphasic} = K_w \frac{(1 + \alpha P_B)}{(1 + \alpha P_A)} \tag{12}$$

i.e. a is the ratio of the volumes of the organic and water phases.

$$\alpha = \frac{V_{org}}{V_w} \tag{13}$$

The apparent equilibrium constant (Kbiphasic) varies with the relative volumes of the aqueous and organic phases; increasing when the substrate is partitioned more efficiently out of the organic phase and into the aqueous phase relative to the behaviour of the product (i.e. PA < PB), and decreasing when the reverse occurs (Figure 3A). If there are sufficient differences in the partition coefficients and both substrate and product are relatively non-polar then substantial shifts in the equilibria occur even at high relative water contents.

A substantial increase in the apparent equilibrium constant may be achieved, when both the substrate and product have partition coefficients less than unity, when the ratio of the partition coefficients is suitable (Figure 3B). Clearly if a is very small $K_{biphasic}$ tends to K_w, the equilibrium constant in aqueous solution. However, if a is sufficiently large, equation 12 simplifies to

$$K_{biphasic} = K_w \frac{P_B}{P_A} \tag{14}$$

Therefore, substituting from equations 2 - 5,

$$K_{biphasic} = \frac{[B_{org}]}{[A_{org}]} = K_{org} \tag{15}$$

Therefore $K_{biphasic}$ tends to K_{org}, the equilibrium constant of the reaction in the pure organic liquid.

In figure 3, (A) shows the effect of the ratio of the partition coefficients, P_B/P_A. The curves have been generated using equation 12 and the following values for the partition coefficients, from the top downwards; P_A = 0.2, P_B = 20; P_A = 0.2, P_B = 2; P_A = 0.2, PB = 0.6; P_A = 0.6, P_B = 0.2; P_A = 2, P_B = 0.2; P_A = 20, P_B = 0.2. (B) shows the effect of the polarity of A and B, keeping the ratio of the

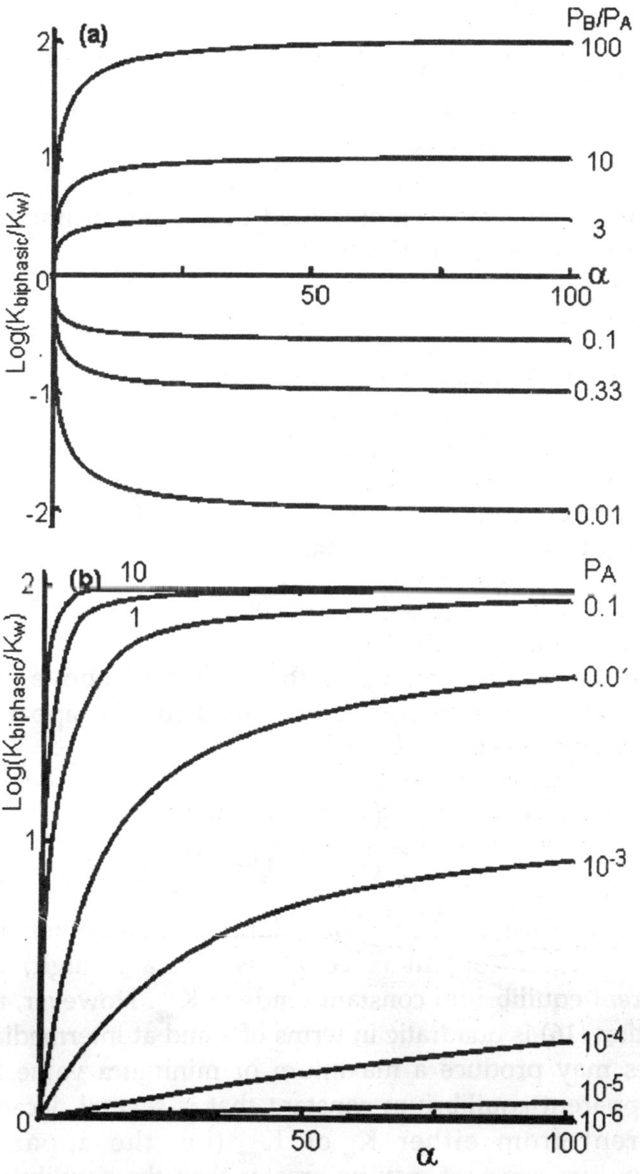

Figure 3. The variation in the apparent equilibrium constants of a one-substrate one-product reaction (A ⇌ B) with the relative component composition of a biphasic aqueous-organic system.

partition coefficients constant ($P_B/P_A = 100$). The following values for the partition coefficients have been used, from the top downwards; $P_B = 100$, $P_B = 10$, $P_B = 1$, $P_B = 0.1$, $P_B = 0.01$, $P_B = 0.001$, $P_B = 0.0001$.

Many reactions are bimolecular, and these are influenced in a more complex way by the biphasic system. The processes may be represented in a manner similar to that used for monomolecular reactions;

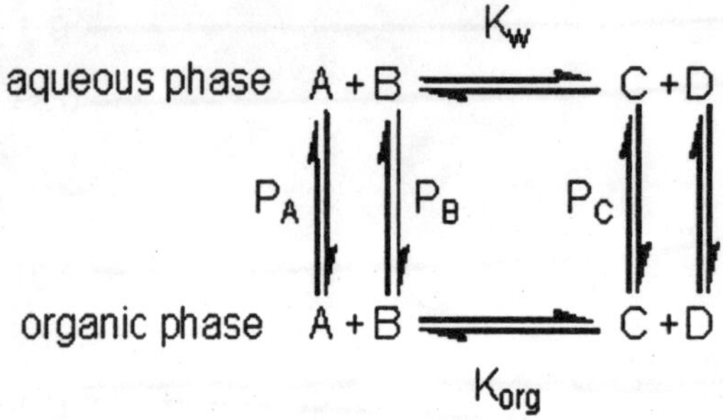

(3)

Using similar reasoning to that outlined above, the following expression may be obtained for the apparent equilibrium constantes

$$K_{biphasic} = K_w \frac{(1+\alpha P_C)(1+\alpha P_D)}{(1+\alpha P_A)(1+\alpha P_B)} \qquad (16)$$

As in the case of monomolecular reactions, when the relative water content is very low (i.e. a is high) the apparent equilibrium constant tends to K_{org}. However, the equation (16) is quadratic in terms of a and at intermediate values may produce a maximum or minimum value for the apparent equilibrium constant that is several decades different from either K_w or K_{org} (i.e. the apparent equilibrium constant may be greater than the equilibrium constant of the same reaction in either of the pure phases, see Figure 4).

Under some circumstances this may enable the reaction productivity in the biphasic system to be much greater than that attainable in either pure phase.

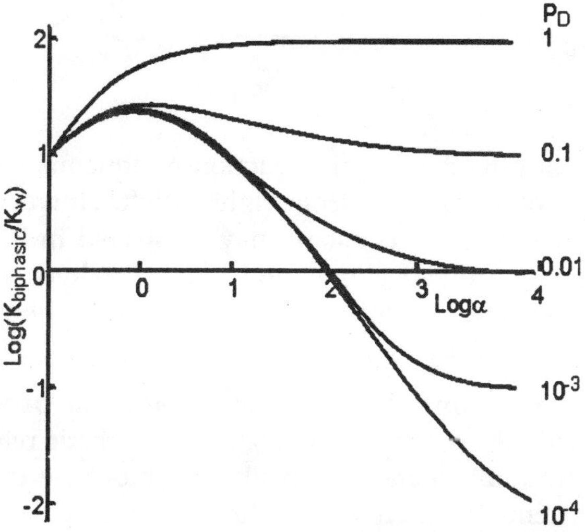

Figure 4. The variation in the apparent equilibrium constants of a two-substrate two-product reaction (A + B [illegible] C + D) with the relative component composition of a biphasic aqueous-organic system.

In all cases the partition coefficients are $P_A = 1$, $P_B = 1$ and $P_C = 100$. The following values for the partition coefficient (P_D) have been used, from the top downwards; 1, 0.1, 0.01, 0.001 and 0.0001.

A special and relevant case of the bimolecular reaction scheme is where one of the reactants is water (e.g. use of the hydrolases). These reactions may be written in a way corresponding to reaction scheme.

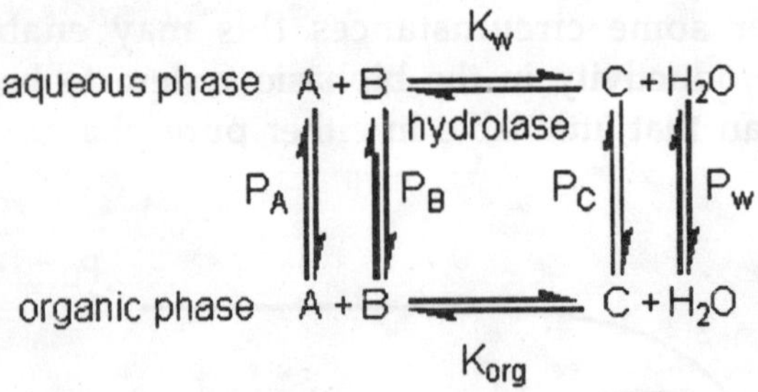

(4)

The normal direction for the reaction in aqueous solution is the hydrolytic process from right to left. However, the apparent equilibrium constant may be shifted in biphasic solutions, as outlined above, and allow the hydrolase to act in a synthetic (i.e. left to right in reaction scheme (4)), rather than hydrolytic, manner.

Such reactions significantly extend the processes available to the enzyme technologist, as synthetic reactions are generally much more difficult to achieve by classical organic chemistry than hydrolytic reactions, particularly when a regiospecificity is required as, then, there is a choice of reactive groups in the substrates. Two factors affect the yield of the product (C) in such reactions: (1) the shift in apparent equilibrium constant ($K_{biphasic}$); and (2) the concentration of water ($[(H_2O)_t]$), as one of the participating reactants. $[(H_2O)_t]$ may be obtained from the material balance,

$$[(H_2O)_t](V_w + V_{org}) = [(H_2O)_w]V_w + [(H_2O)_{org}]V_{org} \quad (17)$$

where $[(H_2O)_w]$ is the molar concentration of pure water (= 1000/18 = 55.5 M). Substituting Pw for $[(H_2O)_{org}]/[(H_2O)_w]$,

$$[(H_2O)_t] = \frac{55.5(1+\alpha P_w)}{1+\alpha} \quad (18)$$

$$K_{biphasic} = \frac{[A_t][B_t]}{[C_t][(H_2O)_t]} \tag{19}$$

Substituting for Kbiphasic from equation 16 (with P_D set to equal P_w) and rearranging, the terms in $(1 + aP_w)$ cancel to give

$$[C_t] = K_w \frac{(1+\alpha P_C)(1+\alpha)[A_t][B_t]}{55.5(1+\alpha P_A)(1+\alpha P_B)} \tag{20}$$

This represents a fairly complex relationship but a few generalisations may be made. If the product (C) is less polar than either reactant (A or B) then the yield generally increases with the relative concentration of the organic phase (a) to reach a plateau. If the reverse occurs, then there will be a plateau at low yield and high a in the a-yield diagram. Both minima and maxima may occur, dependent on the parameters. Figure 5 shows the variation of the apparent equilibrium constant and the yield for the enzymic synthesis of an ester from its acid and alcohol in a biphasic system.

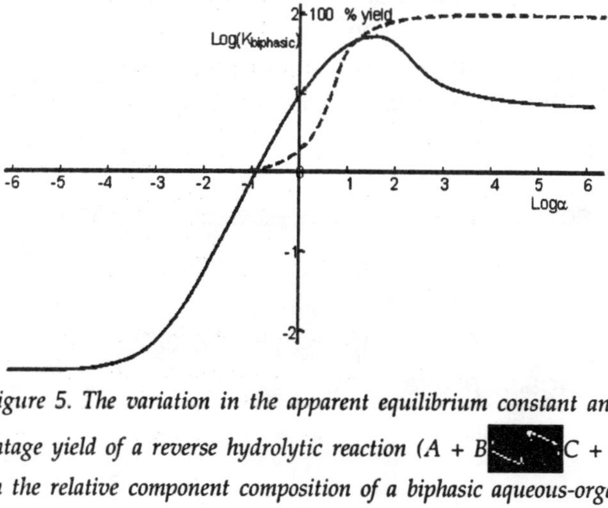

Figure 5. The variation in the apparent equilibrium constant and percentage yield of a reverse hydrolytic reaction $(A + B \rightleftharpoons C + H_2O)$ with the relative component composition of a biphasic aqueous-organic system.

The reaction portrayed is the synthesis of N-benzoyl-L-phenylalanine ethyl ester, from N-benzoyl-L-phenylalanine and ethyl alcohol, catalysed by a-chymotrypsin in a biphasic chloroform-water system (pH 7). --- apparent equilibrium constant; ------- % yield (semi-logarithmic plot). The curves were calculated assuming Kw = 0.002, PA = 0.11 (N-benzoyl-L-phenylalanine at pH 7), PB = 0.01 (ethyl alcohol), PC = 4100 (N-benzoyl-L-phenylalanine ethyl ester), and an equimolar mixture of reactants.

From this it can be seen that the use of a biphasic system can increase the yield of such reactions from zero to nearly 100%. In order to achieve high conversions, the water produced must be removed from the two-phase system. If it is not removed, there is a continual drop in a which reduces the extent of the favourable equilibrium. There are no simple ways in which this water may be removed but reactors which allow the constant addition of fresh catalyst and organic phase reduce the size of the problem.

Use of biphasic solvent systems also affect the ionisation of acid and basic groups. Consider the ionisation of an acid H_A, where only the unionised form is soluble in the organic phase and the ionic species are only soluble in the aqueous phase,

$$\begin{array}{lcl} \text{aqueous phase} & HA & \xrightleftharpoons{K_{a,w}} H^+ + A^- \\ & P_{HA} \updownarrow & \\ \text{organic phase} & HA & \end{array} \qquad (5)$$

$$K_{a,w} = \frac{[H^+][A_w^-]}{[HA_w]} \tag{21}$$

$$pK_{a,w} = pH + Log_{10}\left(\frac{[HA_w]}{[A_w^-]}\right) \tag{22}$$

The hydrogen ion concentration, as determined, is characteristic of the aqueous phase,

$$pK_{a,biphasic} = pH + Log_{10}\left(\frac{[HA_t]}{[A_t^-]}\right) \tag{23}$$

A consideration of material mass balances for A^- and H_A gives

$$[A_t^-]V_t = [A_w^-]V_w \tag{24}$$

$$[HA_t]V_t = [HA_w]V_w + [HA_{org}]V_{org} \tag{25}$$

Substituting for pH from equation 22 into equation 23 gives

$$pK_{a,biphasic} = pK_{a,w} + Log_{10}\left\{\left(\frac{[HA_t]}{[A_t^-]}\right)\left(\frac{[A_w^-]}{[HA_w]}\right)\right\} \tag{26}$$

Substituting from equations 24 and 25 this simplifies to give

$$pK_{a,biphasic} = pK_{a,w} + Log_{10}(1 + \alpha P_{HA}) \tag{27}$$

Similarly, for the protonation of a base

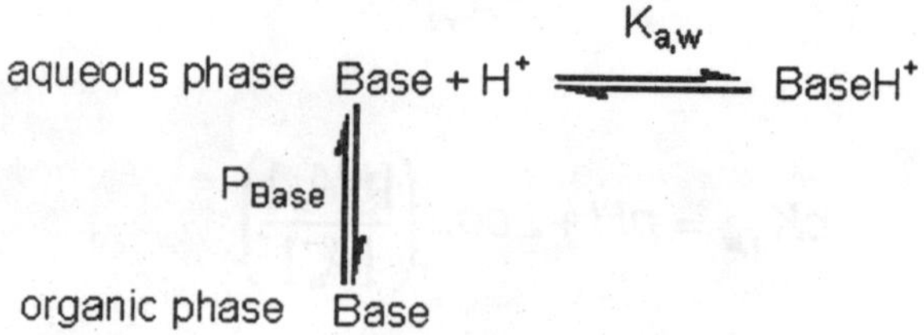

(6)

$$pK_{a,biphasic} = pK_{a,w} - Log_{10}(1+\alpha P_{Base}) \tag{28}$$

A qualitative assessment of reaction schemes (5) and (6) shows that increasing the partition coefficient for the uncharged acid or base will pull the reaction away from the formation of the ionised species. It follows from equations 27 and 28 that the apparent pK_as of acids increase in biphasic systems whereas those of bases decrease. These changes may well be several pH units, dependent on the partition coefficients and the relative fraction of organic solvent present. The shifts in pK_a are in addition to any increase in the pK_a due to the lower dielectric constant of the aqueous phase. This is likely to be particularly relevant under almost anhydrous conditions (i.e. where a is high) which tend to 'freeze' the hydrogen ions lowering their activity. The utility of the shift in the pK_a can be shown using ester synthesis catalysed by α-chymotrypsin as an example.

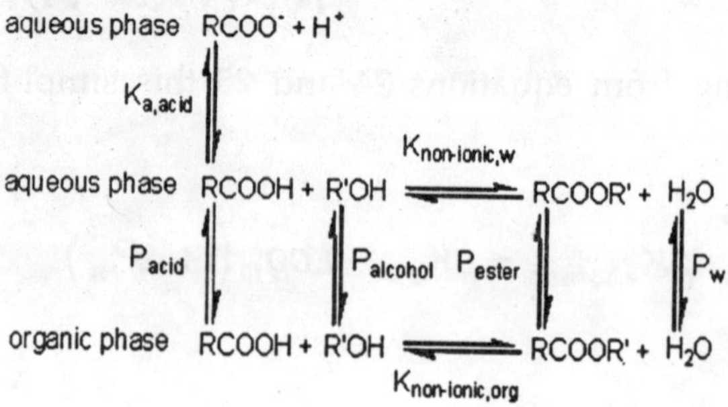

(7)

where RCOOH and RCOO- represent the un-ionised and ionised forms of N-benzoyl-L-phenylalanine, RCOOR' represents N-benzoyl-L-phenylalanine ethyl ester and R'OH represents ethyl alcohol.Although the non-ionic reaction in aqueous solution is evenly balanced (Knonionic,w = 7), the overall reaction proceeds towards the left, in water at neutral pH, due to the pull exerted by the acid's ionisation (pK_a = 3.4). In biphasic chloroform-water solution (a = 20, PHA = 100), the K_a shifts by three decades to give a pK_a close to 7.

This, together with a shift in the equilibrium constant due to the partition effects, outlined earlier, produces an overall shift in the equilibrium constant of about five decades at the optimal pH (5) of the enzyme (Figure 6). Use of the biphasic system allows reactions to be performed at a pH that is both thermodynamically favourable to the reaction direction required and at the optimum for the enzyme's catalytic efficiency.

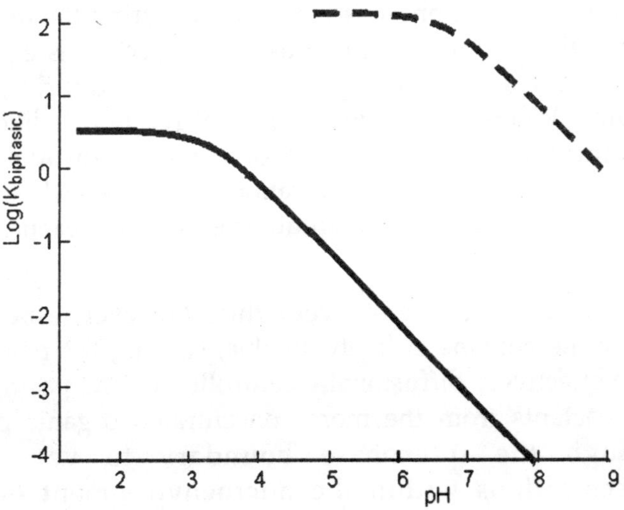

Figure 6. pH dependence of the apparent equilibrium constant for the synthesis of N-benzoyl-L-phenylalanine ethyl ester, as given by the reaction scheme [7]. --- reaction in aqueous solution only (i.e. a = 0); ---------- reaction in a biphasic chloroform-water system (a = 20).

The shift upwards is due to the more significant partition of the ester into the organic phase relative to the reactants, whereas the shift in the pH dependence to higher pH is due to the change in the apparent pK_a of the acid.

Enzyme Kinetics

Although the equilibrium position of reactions may be shifted in biphasic systems, this is of no practical consequence unless the reaction actually occurs. The enzyme must be able to convert the reactants to products. For this to proceed, even in the presence of fully active enzymes, there must be some reactant present in the aqueous microenvironment surrounding the enzyme. Reactants may have fairly low aqueous solubilities under the normal conditions associated with such biphasic processes.

The aqueous phase remains effectively unstirred, even when the organic phase is well-stirred, during the reaction so that the thickness of the 'unstirred layer' (δ) is equal to the thickness of the aqueous layer surrounding the enzyme. Where δ is large, the rate of reaction is likely to be controlled by the passage of reactants through this aqueous layer as the concentration gradients will be very shallow due to the low solubilities of the reactants in water.

However, δ is often very thin, however, especially where the enzyme is freely soluble, causing the reactions to be effectively diffusionally controlled by the passage of the reactants from the more concentrated organic phase through the interphase boundary to very low concentrations within the microenvironment of the enzyme. Clearly, the thickness of the aqueous layer around an immobilised enzyme will be of great importance in determining the degree of diffusional resistance involved, and hence the lowering of the reaction rate.

As with all immobilised enzymes, the volumetric surface area (A/V) is an important parameter governing the overall flux of the substrate to the biocatalytic surface, the effectiveness of the enzyme and the productivity. Whereas in reactions involving a single liquid phase this surface area is the area presented by the exterior of the particles to the bulk of the liquid, in most practical biphasic liquid systems it is the interfacial area between the two liquid phases that is relevant.

Under normal operating conditions this interfacial area will depend upon the volume of the aqueous droplet surrounding and enclosing the enzymic biocatalyst. As the relative volume of the organic phase increases the volumetric surface area will decrease, so reducing the maximum volumetric productivity. This will be of particular importance where a needs to be very high in order to shift the reaction equilibrium.

Clearly this will reduce the reaction rate due to the reduction in the amount of enzyme that can be present, due to its associated water, and the associated reduction in the interfacial area. Where the catalysed reaction generates water, this should not generally be allowed to accumulate as it not only has a detrimental effect on the equilibrium constant but will also slow down the rate of reaction.

There are some rules which allow the optimisation of the two-phase system for enzyme-catalysed reactions. They depend upon the concept of LogP being extended to cover the substrates, products and the interphase (Figure 1).

1. The difference between the LogP values of the substrate(s) and the interphase should be as small as

possible whereas that of the substrate(s) should be much lower than that of the organic phase. These conditions encourage high concentrations of substrate(s) within the interphase and, hence, the transfer of the substrate from the organic phase into the aqueous phase.

2. The difference between the LogP values of the product(s) and the organic phase should be as small as possible whereas that of the product(s) should be much greater than that of the interphase. These conditions encourage the transfer of the product from the aqueous phase through the interphase into the organic phase, after reaction.

The LogP of the interphase can be varied independently of the organic phase by suitable choice of surfactants and cosurfactants. The precise structure of this interphase may be much more complex than outlined in Figure 1. In particular, when there is excess water and surfactant present, the surfactant may form multiple concentric membranes, consisting of surfactant bilayers, around the aqueous core. This presents another reason for the removal of the water produced by use of hydrolases (used as synthetases) in biphasic systems.

Control over the amount of water surrounding the enzyme is often possible by means of molecular sieves (e.g. potassium aluminium silicate), which absorb water. Removal of water dissolved in the organic phase can easily be achieved by their use and this leads to the depletion of water from the aqueous biocatalytic pools by its partition.

If the organic phase has a high boiling point the removal of water may be simply achieved by vacuum distillation; e.g. in the esterification of fatty acids and fatty alcohols catalysed by lipase and used in wax manufacture.

Aqueous Two-phase Systems

Aqueous 2-phase systems offer the opportunity to shift reaction equilibria towards product formation by ensuring that enzyme and substrate partition into one phase whilst the product enters, and may be removed from, the other. The theory developed (earlier) for aqueous/organic biphasic systems is equally applicable to these systems. An example of such an extractive bioconversion is a method for the conversion of starch to glucose using bacterial α-amylase and glucoamylase.

Starch substrate partitions almost entirely into the lower, more hydrophilic, dextran-rich phase of a system comprising 3% polyethylene glycol and 5% crude unfractionated dextran. The enzymes also partition largely to the bottom phase but the glucose, produced by the hydrolysis, distributes itself more evenly between the phases. A small proportion of the glucoamylase enters the upper phase and will convert any oligosaccharide entering that phase to glucose. Concentrations of up to 140 g l-1 glucose may be reached in the upper phase.

Such a system offers some of the advantages of an immobilised enzyme process without some of the disadvantages. The enzymes are largely retained and are stabilised by the presence of the polymers yet catalysis is in homogeneous solution (within the phase) so no diffusion limitations to mass transfer exist. Drawbacks include the need to separate the product from the upper phase polymers and gradual loss of enzymes which enter the upper phase. Enzyme loss could be reduced, without introducing diffusion limitations, by linking them to hydrophilic polymers so as to form soluble complexes.

Use of Enzymes 'in Reverse'

It has long been known that if proteases are supplied with high concentrations of soluble proteins, peptides or amino acids, polymers (plasteins) are produced with apparently random, if rather hydrophobic, structures. This reaction has been used, for example, to produce bland-tasting, colourless plasteins from brightly coloured, unpleasant tasting, algal biomass and for the introduction of extra methionine into low quality soy protein.

However in general, the non-specific use of proteases in the synthesis of new structures has not found commercial use. However, proteases have come into use as alternatives to chemical methods for the synthesis of peptides of known and predetermined structure because their specificity allows reactions to proceed stereospecifically and without costly protection of side-chains.

A method for the synthesis of the high intensity sweetener aspartame exemplifies the power of proteases (in this case, thermolysin). Aspartame is the dipeptide of L-aspartic acid with the methyl ester of L-phenylalanine (α-L-aspartyl-L- phenylalanyl-O-methyl ester). The chemical synthesis of aspartame requires protection of both the β-carboxyl group and the α-amino group of the L-aspartic acid. Even then, it produces aspartame in low yield and at high cost. If the β-carboxyl group is not protected, a cost saving is achieved but about 30% of the β-isomer is formed and has, subsequently, to be removed.

When thermolysin is used to catalyse aspartame production the regiospecificity of the enzyme eliminates the need to protect this β-carboxyl group but the α-amino group must still be protected (usually by means of reaction with benzyl chloroformate to form the benzyloxycarbonyl derivative, i.e. BOC-L-aspartic acid) to prevent the synthesis of poly-L-aspartic acid. More economical racemic amino acids can also be used as only the desired isomer of aspartame will be formed.

If stoichiometric quantities of BOC-L-aspartic acid and L-phenylalanine methyl ester are reacted in the presence of thermolysin, an equilibrium reaction mixture is produced giving relatively small yields of BOC-aspartame. However, if two equivalents of the phenylalanine methyl ester are used, an insoluble addition complex forms in high yield at concentrations above 1 M. The loss of product from the liquid phase due to this precipitation greatly increases the overall yield of this process. Later, the BOC-aspartame may be released from this adduct by simply altering the pH.

The stereospecificity of the thermolysin determines that only the L-isomer of phenylalanine methyl ester reacts but the addition product is formed equally well from both the D- and L-isomers. This fortuitous state of affairs allows the use of racemic phenylalanine methyl ester, the L-isomer being converted to the aspartame derivative and the D-isomer forming the insoluble complex shifting the equilibrium to product formation.

D-phenylalanine ethyl ester released from the addition complex may be isomerised enzymically to reform the racemic mixture. The BOC-aspartame may be deprotected by a simple hydrogenation process to form aspartame.

BOC-L-aspartic acid + L-phenylalanine methyl ester

⇅ thermolysin

BOC-L-aspartame

↓ D-phenylalanine methyl ester (8)

BOC-L-aspartame + D-phenylalanine methyl ester adduct
(precipitates)

Immobilised thermolysin cannot be used in this process as it has been found to co-precipitate with the insoluble adduct. However, this may be circumvented by its use within a liquid/liquid biphasic system.

The synthesis of aspartame is a very simple example of how proteases may be used in peptide synthesis. Most proteases show specificity in their cleavage sites and may be used to synthesise specific peptide linkages. Factors that favour peptide synthesis are correct choice of pH, the selection of protecting residues for amino and carboxyl groups that favour product precipitation and the use of liquid/liquid biphasic systems, all of which act by controlling the equilibrium of the reaction.

An alternative strategy is kinetically controlled synthesis where the rate of peptide product synthesis (k_P) is high compared with the rate of peptide hydrolysis (k_H). This may be ensured by providing an amino acid or peptide which is a more powerful nucleophile than water in accepting a peptide unit from an enzyme-peptide intermediate. This kinetically controlled reaction may be represented as

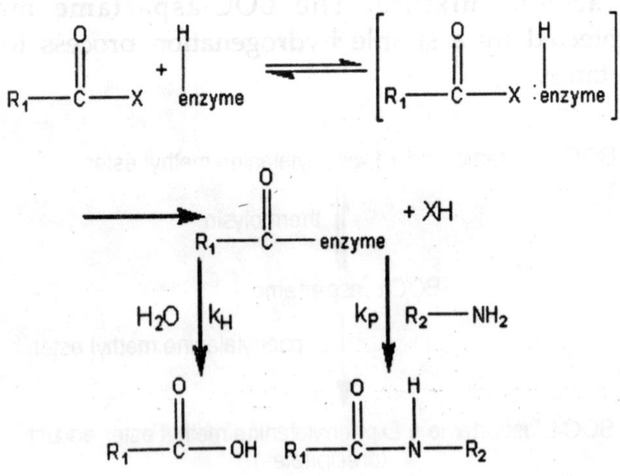

(9)

where X represents an alcohol, amine or other activating group, i.e. the reactant is an ester, amide (peptide) or activated carboxylic acid. The relative rate of peptide formation compared with hydrolysis depends on the the ratio k_P/k_H, the ratio of the K_m values for water and amine, and the relative concentrations of the (unprotonated) amine and water. Thus, where necessary, the reaction yield may be improved by lowering the water activity.

It has been found that the yields may be increased by reducing the temperature to 4°C, perhaps by a disproportionate effect on the K_m values. The amine and enzyme concentrations should be as high as possible for such kinetically controlled reactions and only those enzymes which utilise a covalently-linked enzyme-peptide intermediates can be used.

Also, the reaction is stopped well before equilibrium is reached as, under such thermodynamic control, the product peptide will be converted back though the enzyme intermediate to the carboxylic acid, a process made almost irreversible by its ionisation in solutions of pH above its pK_a, as shown in reaction scheme (10).

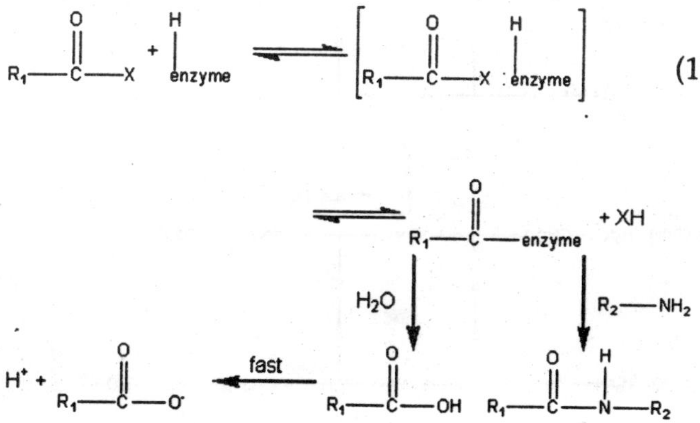

(10)

Peptides may be lengthened either by the addition of single amino acid residues or by the condensation of peptide fragments. The size of the peptide required as the final product may determine the type of synthesis used; condensation of fragments may be performed in kinetically controlled processes whereas stepwise elongation is best achieved using biphasic solid/liquid (i.e. precipitation) or liquid/liquid thermodynamically controlled processes.

In general, few proteases are required for such synthetic purposes. As they are quite costly, especially at the high activities necessary for kinetic control, they may be immobilised in order to enable their repeated use. Many examples of enzymic peptide synthesis may be sited. The conversion of porcine insulin to human insulin requires the replacement of the C-terminal alanine (b30) residue by a threonine. This can be achieved by a single transpeptidation step catalysed by trypsin using an carboxyl-protected threonine in an aqueous solution with an organic co-solvent. The protective group can be simply removed later by mild hydrolysis and the product purified by silica gel chromatography.

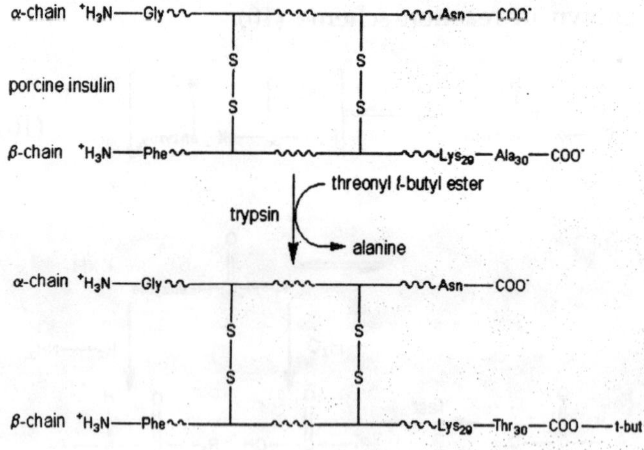

(11)

Contd . . .

Glycosidases used in Synthetic Reactions

This is a comparatively neglected topic probably because polysaccharides can be obtained readily from plant or microbial sources, because polysaccharide function is not so specifically related to its structure, and because the theory describing the functional significance of oligosaccharides is still being developed.

As there is an abundance of knowledge about glycosidases and their specificities, there is no fundamental reason why they should not be used to synthesise oligosaccharides. There are few examples in the literature of oligosaccharides being constructed using enzymes such as dextransucrase (EC 2.4.1.5), levansucrase (EC 2.4.1.10), cyclomaltodextrin glucanotransferase (EC 2.4.1.19) and b-galactosidase (EC 3.2.1.23) to build up glucose, fructose or galactose residues on suitable acceptors.

Fructosyltransferases (e.g. inulosucrase, EC 2.4.1.9) have been used to synthesise compounds such as sucrose-6-acetate, xylosucrose (β-D-fructofuranosyl-(2,1)-α-D-xylopyranoside) and the low cariogenic sweetener, 'neosugar' which consists of a mixture of glucose, sucrose and β-2,1 linked fructans with terminal non-reducing glucose residues (typical composition before chromatographic refinement: 2% fructose, 26% glucose, 11% sucrose, 30% 1-kestose (β-D-fructofuranosyl-(2,1)-β-D-fructofuranosyl- (2,1)-α-D-glucopyranose), 25% nystose (β-D-fructofuranosyl- (2,1)-β-D-fructofuranosyl-(2,1)-β-D-fructofuranosyl-(2,1)-α-D- glucopyranose) and 6% higher oligosaccharides).

$$\text{sucrose + glucose-6-acetate} \rightarrow \text{sucrose-6-acetate + glucose} \quad (12)$$

$$\text{sucrose + xylose} \rightarrow \text{xylosucrose + glucose} \quad (13)$$

$$\text{sucrose + sucrose} \rightleftharpoons \text{1-kestose + glucose} \quad (14)$$

$$\text{1-kestose + sucrose} \rightleftharpoons \text{nystose + glucose} \quad (15)$$

As shown in reaction scheme [16], dextransucrase can be used to produce dextran (a polymer of a-1,6 linked glucose residues) from sucrose. Dextran is used in gel chromatographic media, such as Sephadex, in the aqueous 2-phase systems already described, and as a blood plasma extender. For any of these uses it is undesirable to have cells remaining in the dextran so the dextransucrase, produced extracellularly by Leuconostoc mesenteroides is purified before use.

The size of dextran molecules produced may be controlled by including small concentrations of sugars such as maltose which compete with sucrose as acceptors for glucose residues from the donor sucrose molecules.

$$\text{sucrose + dextran (Gn)} \rightleftharpoons \text{fructose + dextran (Gn+1)} \quad (16)$$

Thermodynamically controlled synthesis of oligosaccharides, catalysed by glycosidases, is possible by use of high substrate concentrations and a 'molecular trap' to remove the products. For example, β-galactosidase may be used to synthesise N-acetyllactosamine (β-D-galactopyranosyl-(1,6)-2-acetamido- 2-deoxy-D-glucose) using a carbon-celite column to remove it, as formed.

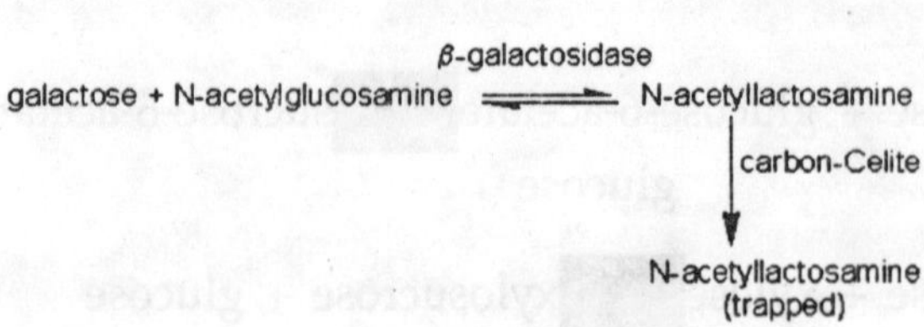

(17)

There seems to have been no systematic attempt to construct hetero-oligosaccharides in a stepwise movement yet thermodynamically or kinetically controlled reactions could be used just as in peptide synthesis.

Interesterification of Lipids

There is no opportunity or established need to build up polymers or oligomers using lipases or esterases yet it is possible and commercially advantageous to use these enzymes as transferases in transesterifications (carboxyl group exchange between esters), acidolyses (carboxyl group exchange between esters and carboxylic acids) and alcoholyses (alcohol exchange between esters and alcohols).

Certain triglycerides, cocoa butter being the outstanding example, have high value because of their physical properties and comparative rarity. There is a commercial pull, therefore, to fund routes to the production of high value triglycerides from more plentiful, cheaper raw materials. This may be done using lipases acting as transacylases.

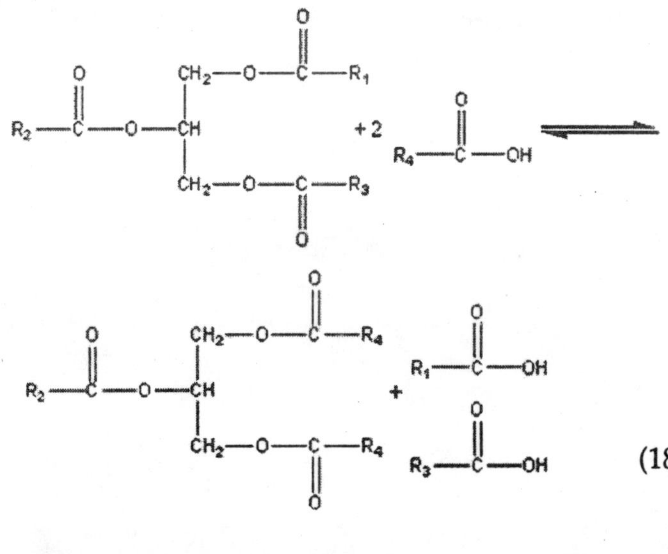

(18)

This acidolysis reaction may be used for increasing the value of rapeseed oil by exchanging linoleic acid for linolenic acid residues and increasing the value of palm oil and sunflower oil by increasing their content of oleic acid residues. Cocoa butter is a relatively expensive fat, used in confectionery, because of its sharp melting point between room temperature and body temperature; chocolate literally melts in the mouth.

This is due to the fairly small variation in the structure of the constituent triglycerides; 80% have palmitic acid or stearic acid in the 1 and 3 positions with oleic acid in the central 2 position. For the production of cocoa butter substitute from palm oil, a process which increases the value of the product three-fold, the acidolysis utilises stearic acid (i.e. R4 is C17H35) in hexane containing just sufficient water to activate the lipase.

Olive oil may be similarly improved by exchanging its 1,3-oleic acid residues for palmityl groups. The products may be recovered by recrystallisation from aqueous acetone. Such reactions may also be used for the resolution of racemic mixtures of carboxylic acids or alcohols, the lipase generally being specific for only one out of a pair of optical isomers.

11

Challenges and Opportunities in Enzyme Manufacturing

Bioprocessing offers a level of specificity, predictability, and productivity that otherwise would not exist in the manufacture of these products. Moreover, when the raw materials contain certain molecules with complex structures, bioprocesses enable the synthesis of products that cannot be made by any other means. Combined, these capabilities provide for new process designs that are cost effective, energy efficient, product specific, and environmentally benign.

While bioprocessing already is employed in industry, research is needed to define the parameters of the related processes precisely in order to expand use of this approach and derive the maximum possible benefits. In US, the Federal Government plays a critical role in this area because individual manufacturers are unlikely to invest in time-consuming generic research for the benefit of entire industries.

Process design involves the integration and scale-up of a number of operations, including upstream and downstream processing, and process monitoring, optimization, and control. One objective for process design is to quantify process variables and product yields in order to obtain a cost estimate for producing a material of biological origin, and to evaluate alternative designs that minimize the cost of producing the product. An optimal process can be one that minimizes the cost of producing a material, maximizes the profit, or assures that an operation meets special safety or environmental requirements.

Upstream Processing

Upstream processing encompasses any technology that leads to the synthesis of a product as well as the fundamental science and engineering needed to understand product formation. Specific areas of upstreaming process include biocatalysis, metabolic engineering, biomass conversion, bioreactor design and cell culturing techniques, and transgenic animals.

Biocatalysis

Nature's catalysts — enzymes — function highly selectively to accelerate the rate of chemical reactions in biological systems. Enzymes also can limit reactivity toward a particular substrate and limit chemical conversion to a single desired product. For many substrates, enzyme-catalyzed reactions exhibit rates that are many orders of magnitude faster than those of uncatalyzed reactions.

In chemical synthesis and other processes of industrial importance, the ability to mimic the selectivity and rate-enhancing properties of enzymes would be of substantial benefit in reducing economic and environmental costs. In some cases, this approach might provide a commercial means of synthesizing a material where only much less desirable methods were available previously.

At present, about 50 enzymes are used in industry. Most of these industrial uses involve breaking down large molecules into simpler ones. Examples include the proteases in laundry detergents, and the hydrolases used in starch processing. The commercial benefits of enzymology would be enhanced if technologists could expand capabilities for creating complex molecules from simpler ones or transforming existing chemical structures into more active compounds. Production of sweetness-enhanced corn syrup by the enzyme glucose isomerase is one noteworthy example of how a chemical transformation can be carried out on the industrial scale. Synthesis routes that can distinguish between right- and left-handed chemical structures are being pursued for pharmaceutical, food, and agricultural applications. Many industrial firms are exploring the unique capabilities of enzymes to carry out such chemical transformations.

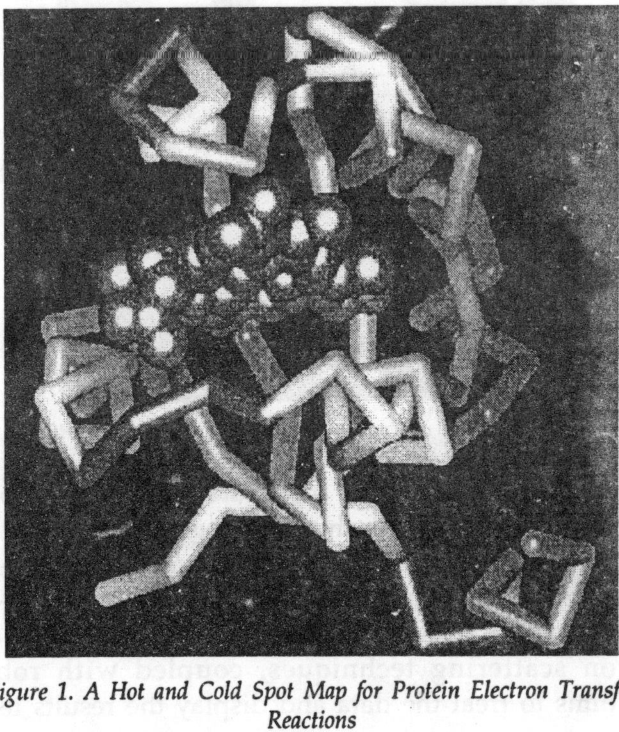

Figure 1. A Hot and Cold Spot Map for Protein Electron Transfer Reactions

Many research efforts are under way, and solid progress has been made using protein engineering, chemical modification, and recombinant monoclonal antibody technology. More recently, recombinant DNA technology has been used to generate catalysts that exhibit the useful characteristics of naturally occurring enzymes but have been designed either to catalyze chemical reactions not found in nature or to alter biological reactions to broaden their specificity and expand their potential for practical application.

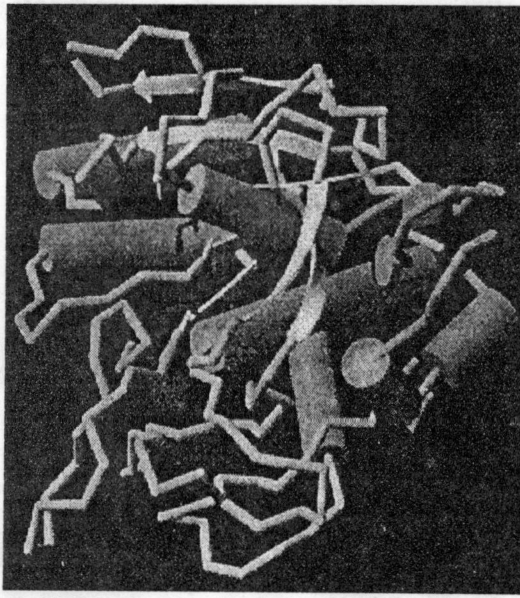

Figure 2. Structure of Engineered Detergent Enzyme Subtilisin

Progress in this field will accelerate with an improved understanding of relationships among protein structure, function, and energetics. More efficient methods are being developed to obtain protein structure using X-ray crystallography, nuclear magnetic resonance (NMR), and neutron scattering techniques, coupled with robust algorithms to treat the data and display the results using computer graphics. The use of computer graphics to visualize the dynamics of proteins also will enhance understanding of the mechanisms of enzyme action.

Progress in obtaining structural data in real time ultimately will help modelers develop algorithms to predict the dynamics of enzymes. Research also should be supported on the thermodynamics and kinetics of prototypical enzyme-catalyzed reactions of interest to the bulk and commodity chemical markets; examples include the reactions involved in corn processing and the action of laundry detergents.

Metabolic Engineering

Metabolic engineering is the use of recombinant DNA technology to enhance the activities of a cell by manipulating its metabolic pathways. To exploit this approach, scientists must develop an improved understanding of cell metabolism. Experimental and mathematical techniques are needed for quantifying the effects of altered uses of raw materials by cells. Techniques such as electrophoresis can be used to distinguish and quantify cellular proteins produced in response to altered metabolism. In addition, non-invasive techniques such as near-infrared spectroscopy and NMR can quantify intracellular concentrations of raw materials. Also of interest is research on thermodynamics and kinetics of prototypical microbial and microalgal systems.

The first applications of metabolic engineering have incorporated genes for pathway enzymes and transport proteins. In this manner, for example, the bacterium *E. coli* has been induced to produce ethanol. But amplified expression of non-native proteins can affect host cell metabolism — sometimes interfering with the engineering objective — because raw material resources are reallocated. The effects may include a slowing of cell growth and possibly cell death. Often, the expression of small amounts of non-native proteins increases the level of stress proteins.

This problem is ubiquitous, occurring in simple organisms such as *E. coli* and in complex cells, such as yeast cells, human cells, the chinese hamster ovary cell (used commercially for protein synthesis), and an insect cell line that is a possible source of proteins for making AIDS vaccines.

Biomass Conversion

Biomass includes organic polymeric material — such as lignin, starches, celluloses, and oils — produced by biological processes. Plants and algae are the main sources of biomass, generating many billions of tons annually through photosynthesis. Other sources of biomass derived from human activities include food processing wastes, waste paper, and municipal solid wastes.

Biomass is a potential source of energy resources such as methane and ethanol, commodity chemicals, animal feed, and specialty products (e.g., flavors, fragrances, pigments). Biomass is an attractive alternative to petroleum- based sources because it not only is a low-cost substrate for production, but also is environmentally friendly; it is renewable and may reduce emissions during processing and improve the biodegradability of end products.

Biotechnology is especially relevant to the commercial use of biomass because it employs natural processes uniquely suited for creating specific products. An example of such a process is fermentation, which converts glucose sugars derived from biomass into commodity chemicals. The components of plant biomass, and products that can be derived from them, include

— cellulose (a mostly crystalline polymer of glucose molecules), which can be hydrolyzed (i.e., broken down) to form glucose, which in turn can be bioconverted into a variety of products, such as ethanol;

— hemicellulose (another heterogeneous, amorphous sugar polymer), which can be hydrolyzed to make sugars such as xylose and glucose;

— lignin (a complex aromatic polymer), which can serve as a chemical feedstock or be burned to provide heat and electricity;

— starch, used by the corn wet milling industry and fuel alcohol industry as the basis for large, commercial-scale bioprocesses; and

— oils. Production of diesel fuel from oil seeds is probably the most well known industrial application of a "renewable" oil; the hydroxy fatty acids from oil seeds can be used in a wide range of products, including cosmetics, waxes, nylons, plastics, and coatings.

Fuels and chemicals can be produced not only from oil seed crops (e.g., soybeans, canola, and crambe), but also from unicellular algae and waste fats from animals. The substantial fractions of oils in these resources can be transformed through a chemical process that changes the large molecules in fats and oils into smaller molecules.

Bioreactor Design and Cell Culturing Techniques

If the great variety of potentially commercial bioprocesses are to be developed and applied, then bioreactors must be designed in which the environment can be controlled precisely to maximize process efficiency.

The design of a bioreactor requires a basic understanding of both chemical reactor design and cell biology. First, designers must understand the effects of reaction rates and stoichiometry, mass transfer, heat transfer, and turbulence and mixing on product distribution, reactor productivity and size, and operational characteristics. These phenomena need to be expressed in accurate but tractable models that can be used for design and optimization calculations.

In order to develop effective cell culturing techniques, designers also must have basic knowledge of cellular functions and protein chemistry. They should understand the molecular, genetic, and metabolic processes involved in the growth of cells and the expression of cellular products; and structure/function relationships in the use of proteins for biochemical conversions.

Transgenic Animals

Transgenic animals can be employed either as biofactories for the production of commercial products or as living models for the study of human diseases and evaluation of pharmaceuticals. The use of transgenic animals for these purposes can be more economical or, in the case of human disease and drug models, more realistic than are conventional alternatives.

Transgenic animals are being developed for a wide variety of applications. Transgenic cows are used for production of lactoferrin, a naturally occurring milk protein used in products such as baby formula. Several companies plan to market pharmaceuticals from the milk of sheep and goats and the blood of pigs. Anti-clotting factors used to treat cardiovascular diseases have been produced in rabbits, goats, and mice.

Downstream Processing

Downstream processing includes the cost-effective separation and purification of bioproducts, as well as biorefining. Often, downstream processing is the most expensive phase of producing a substance of biological origin, especially for products with stringent regulatory requirements. A key reason for this high cost is the complex and dilute nature of the aqueous solutions in which bioproducts generally are produced; an inverse relationship has been demonstrated between the price of biological products and the strength of the concentrations from which they must be isolated. This is the case for bulk and commodity chemicals, including ethanol from biological sources.

The high cost of downstream processing means there is significant potential for savings from improved processes. For regulated bioproducts, such as pharmaceuticals, removal of trace impurities is the expensive step in the purification.

Separation and Purification

The separation and purification of materials produced in a bioreactor is a critical part of a manufacturing operation. The biological products involved range from high-value-added substances used as pharmaceutical agents (e.g., insulin) to lower-cost products including commodity chemicals (e.g., ethanol).

High-value bioproducts are usually fragile molecules, such as proteins or peptides, that require highly specialized and mild processing conditions and may need to be separated from a complex mixture of molecules, including cell debris. This combination of factors makes separation difficult. At present, most separation schemes are scaled-up laboratory procedures; research is needed to improve their performance.

Biological methods hold promise for improving separations. For example, a cell could be modified genetically to suppress production of undesirable by-products and alter the cell wall so it would be permeable only to the desired product. It is also possible to genetically engineer a product so that it passes through the cell wall, has an affinity for certain separation matrices, or is joined to another molecule with desirable separation characteristics. In addition to improving separations, biological approaches also reduce energy expenses, a major part of the cost of separating biological products.

Biological processes one day may offer economical alternatives to current, petrochemically based methods for manufacturing organic acids and alcohols. However, before these bioprocesses can become commercially viable, nontraditional, lower-cost separation methods need to be developed. Research is under way to develop extracting solvents, resins (separation media), and sorbents that are more selective and have a higher capacity than do current materials. Reversible extraction systems are needed that respond to changes in temperature, pressure, or acidity. Combinations of conventional separation methods and biological methods are being explored to reduce product inhibition, which often occurs in fermentations that produce alcohols and solvents.

In addition, mathematical models of separation steps need to be developed to help reconcile regulatory requirements with basic process conditions during early stages of process development, and to meet the demands of a competitive business environment. Modeling enables scale-up considerations to be estimated very quickly, a capability needed in order to commercialize a bioprocess in an industry where being the "first to market" is a critical element of success.

Biorefining

Biorefining involves the use of microbes in mineral processing systems. Biorefining is environmentally friendly and in some cases enables the recovery of minerals and use of resources that otherwise would not be possible. For example, a significant amount of copper is biorefined from slag heaps. Research now under way or planned addresses the use of microorganisms to bioleach oxide and sulfide ores, and to concentrate metals.

Current research on bioleaching of oxide and sulfide ores addresses treatment of manganese, nickel, cobalt, and precious metal ores. The objective is to identify metabolic pathways in the microorganisms responsible for metal solubilization, and to improve their survival rates and stability. Increased understanding of metabolic pathways will open the door to the manipulation of parameters such as kinetics and metal selectivity, with the aim of enhancing mineral recovery.

An intriguing potential use of biotechnology is *in situ* bioleaching of ore deposits or waste piles. Research is needed to develop a mechanistic understanding of the bioleaching process and to identify environmental and process factors affecting biosystem performance. In addition, bioprocess monitoring schemes and cell-free leaching systems should be developed. Cell-free or cell component systems, in which cellular components rather than live organisms are used for a specific reaction, eliminate the need to provide an environment conducive to survival of a microorganism. Such approaches also tend to enhance control over the reaction, due to the specificity of the cell component.

Biosorption and metal recovery from dilute aqueous solutions is an emerging field of interest, from both a resource conservation standpoint and an environmental remediation standpoint. Microbes, algae, cell wall material, proteins, and other types of biomass have been investigated for use in this application, but economical methods for selective metal recovery have yet to be developed.

Use of microorganisms to remove fine particles from aqueous process streams is also of interest. Laboratory-scale studies have shown that certain microorganisms can cause flocculation of fine mineral suspensions, while others can function as flotation collectors or depressants. Extensive research will be needed before the use of extracellular colloidal molecules (such as polysaccharides and exopolymers released by live biomass) can become a viable alternative to current chemical-based systems.

Process of Monitoring and Control

Process monitoring and control requires knowledge of the current state of every step in a bioprocess and the ability to control the process for optimal results. The most efficient use of upstream and downstream processing depends on knowledge of the state of the system and on control algorithms that can optimize the process and maintain it.

Control and optimization is challenging due to several characteristics of bioprocesses, including their high degree of nonlinearity (meaning that nonlinear differential equations are required for mathematical modeling), especially in batch and fed-batch bioreactors; and their potential for instability when they involve high-yield mutant or recombinant organisms. There are technical obstacles as well, including the unavailability of reliable on-line real-time sensors and realistic models that capture the complexities of biological systems by identifying and quantifying rate-controlling steps in reactions.

Optimization and control methods should be robust, adaptive, and suited to nonlinear processes. Important components of these systems include biosensors and bioelectronics/bionetworks. Because an optimal process often is operating at the limits of process variables, process control can be very difficult; consequently, design methods should consider simultaneously both the economics and the controllability of the process. The affordability of some production processes, particularly for lower-value products, depends on the development of highly accurate sensors and models and the capability to reproduce bioreactor operation precisely.

Biosensors

Sensors are invaluable in the design and operation of automated and environmentally benign manufacturing processes, and in detection, monitoring, and control of food additives, food safety factors, and bioremediation technologies. Biosensors are attractive because they can harness the inherent specificity of many biochemical processes to allow for accurate and highly specific detection of organic and inorganic species in real time. Biotechnology can be used to design, construct, probe, and alter sensitive and highly specific biosensors and incorporate them into transducing and reporting systems.

A biosensor encompasses three major components: a biological component, an interface, and a transducing element. The biological component (e.g., enzyme, immunoprotein, nucleic acids, whole cell) interacts specifically with the substance to be detected. The interface component (e.g., polymeric thick or thin film, chemically modified surface) links the biological component with the transducer. The transducer converts the biochemical interaction into a quantifiable electrical or optical signal.

An example of a biosensor is the plasmon sensor, which consists of a monolayer of immunoproteins immobilized on the surface of a thin film of gold. The analyte — whatever compound binds to the protein employed — attaches to the immunoprotein and changes the properties of surface electromagnetic waves that can be excited in the gold film. This change, which signifies the presence of the analyte, can be detected optically.

A vast store of fundamental information is available on the biological components used in biosensors. The key challenge now is to improve the manufacturing process so that reproducible, dependable, low-cost biosensors can be mass produced.

There are a number of technical obstacles. Poor adhesion of polymer films to the transducer, and hence poor reliability, is a major problem. The film must be compatible with the biological component and also adhere to the transducer surface. Another problem is that insufficient knowledge of methods and effects of immobilizing biological components in/on polymeric films has led to wide variation in sensor response, complicating the manufacture of commercial biosensors. Non-specific binding must be minimized, because the specificity of analyte detection is degraded when other molecules compete for binding sites.

There are also operational difficulties that preclude the use of biosensors in bioprocessing environments. For example, many sterilization techniques (such as heat) either modify or denature biological molecules. There are also problems with miniaturization, sensitivity, dynamic range, and noise reduction related to specific biosensor fabrication strategies. Finally, there is the question of quality control — determining quickly and nondestructively whether the biosensor is operating with prescribed precision and accuracy.

Bioelectronics and Bionetworks

Bioelectronics is an emerging technology that employs biological molecules instead of inorganic materials in conventional integrated-circuit technology or in applications involving unconventional architectures, such as optical processors. The driving force for this research is the possibility of constructing devices on the molecular level and thereby achieving extremely high densities of data storage sites and nano-sized computers.

Biological systems are capable of storing information on the molecular level and processing information along pathways defined on a molecular level. Although biological processes function more slowly than do conventional solid-state devices, this penalty is more then offset by a huge increase in the density of operating units. Various biomimetic or biologically based materials, such as the protein bacteriorhodopsin, are being evaluated for use in bioelectronics.

In living things, data processing is achieved by arrays of neurons. Although the operation of single neurons is well understood, the operation of biological neural networks remains largely unexplored. Recent achievements in the culturing of a monolayer of neurons on a micro-electrode array promise to provide some insight into the operation of neuron arrays. Although neuron devices — bionetworks — would not operate on a molecular scale, they have the potential to form the basis for new computer architectures, including parallel processors.

The major technical obstacle to development of practical bioelectronic devices lies in determining how to preserve and control the properties of the active state when the bioactive species is immobilized in a artificial membrane. Research to explore the fundamental process of biological self- assembly, a capability inherent in biological molecules such as lipids, DNA, and proteins, may be useful in resolving some of the organizational problems. The pattern inherent in a bi-lipid layer or an alkane thiol monolayer ("smart" organic compounds) on a metal surface can be used as a template for further organization of biomolecules.

Biometerials and Processing

The tools of biotechnology can be employed to endow materials with properties not achievable using more conventional means.

Biomolecular Materials

Due to their diversity, versatility, and unusual combinations of properties, biomolecular materials offer promise for application in virtually all sectors of the economy, including defense, energy, agriculture, health, and environmental technology. Examples of biomolecular materials include silk obtained from spiders, and the ceramics in sea shells.

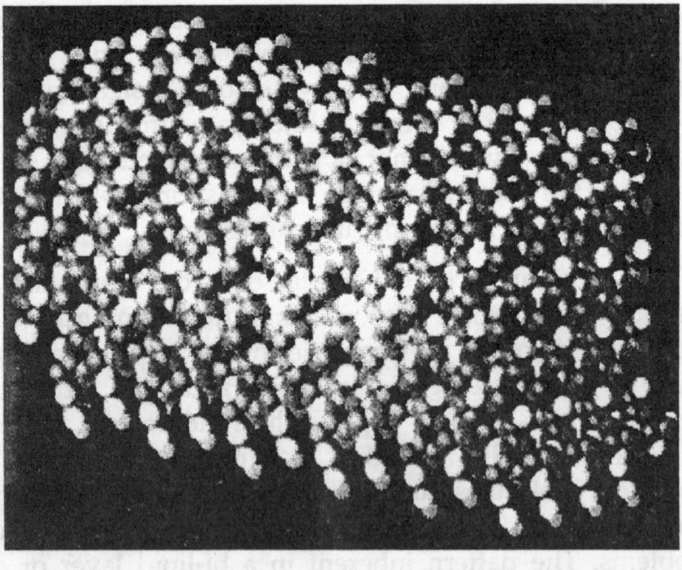

Figure 3. Model of Genetically Engineered Silk-like Protein.

In addition to their direct use as natural cellular products or modified derivatives, biomolecular materials serve another very important purpose by demonstrating how nature has optimized their physical properties. Research is needed to clarify how higher- order structure is achieved and how it serves to determine macromolecular function in such a variety of forms. With continued advances in modern biology, molecular genetics, and protein engineering, and with rapid improvements in physicochemical characterization, novel biomolecular materials can be designed and tailor-made to meet specific needs. This, in turn, would expand the possibilities for practical applications of these materials.

Standard chemical synthesis has inherent limitations, including the production of compounds with unwanted impurities and by- products. By contrast, biomolecular synthesis — manufacturing methods based on biological processes — allows precise control, thereby reducing levels of impurities and by-products. For instance, when a cell produces peptide polymers, control of the amino acid sequence is assured by the fidelity of RNA and DNA replicative mechanisms.

Tissue Engineering

Tissue engineering — a term coined in 1987 referring to the development of biological substitutes to restore, maintain, or improve human tissue function — employs the tools of biotechnology and materials science as well as engineering concepts to explore structure-function relationships in mammalian tissues. This emerging technology could provide for substantial savings in health care costs and major improvements in the quality and length of life for patients with tissue loss or organ failure.

Advances in the study of tissue growth and regeneration, at both the cellular and tissue levels, set the stage for practical application of tissue engineering. The culturing of cells in two-dimensional monolayers enabled the study of cellular processes and opened the door to genetic manipulation. Scientists and engineers have begun to view cell culturing as a three-dimensional process, in which external forces on the cells not only may influence cellular products, but also may reawaken the cellular differentiation process. In order to develop living tissue equivalents, it will be important to understand how the cellular environment affects the differentiation process as well as interactions between the engineered tissue and the host.

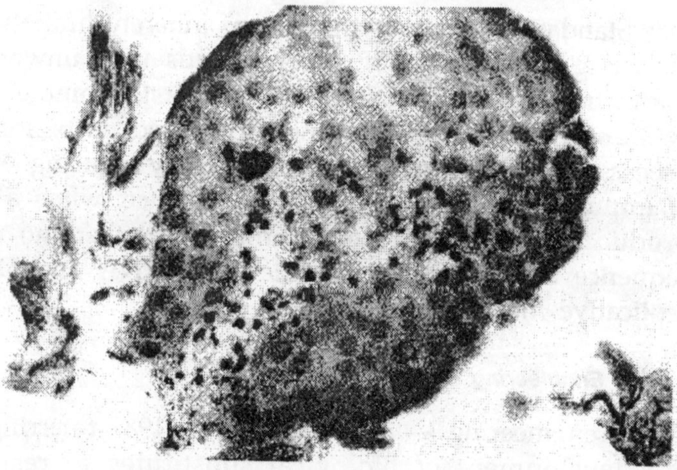

Figure 4. Rotating Tissue Culture Vessels: Growing 3-Dimensional Human Tissues

The first success with differentiated cells came with engineered human skin, now in clinical trials. Scientists also are beginning to explore the potential to grow many tissues in culture. Using stromal cells from human tissue, researchers are developing blood vessels, bone, cartilage, nerve, oral mucosa, bone marrow, liver, and pancreatic cells.

Encapsulated cell therapy is an example of a technique under development by industry that employs biomaterials in the treatment of certain serious, chronic diseases. The goal of this approach is to replace cells within the body that have been destroyed by disease in order to augment circulating or local levels of the deficient molecules.

An encapsulated cell implant consists of cells that secrete the desired hormones, enzymes, or neurotransmitters, enclosed within a polymer capsule and implanted into a specific site within the host. Animal studies have shown that the functional activity of secreting cells can be maintained in vivo. The capsule wall is designed to allow passage of small molecules (i.e., glucose, other nutrients, therapeutic molecules) but prevents or retards the passage of large molecules, such as elements of the immune system. Studies suggest that the transplanted cells are protected from destruction and perhaps even recognition by the host's immune system, allowing the use of unmatched or even genetically altered tissue without systemic immunosuppression.

Thus, the use of an encapsulating membrane may overcome two of the difficulties that prevent widespread tissue transplantation into humans: the limited supply of donor human tissue, and the toxic effects of immunosuppressive drugs required to prevent rejection of unencapsulated transplants.

Bibliography

Bender, M.L., Kezdy, J.F. & Gunter, C.R., "The anatomy of an enzymatic catalysis: a-Chymotrypsin". *Journal of the American Chemical Society*, 86, 3714-21, 1964.

Cornish-Bowden, A., *Principles of enzyme kinetics*. London: Butterworth, 1976.

Fersht, A., *Enzyme structure and mechanism*, 2nd edn, New York: W.H.Freeman & Co.Henderson, P.J.F. (1978). Statistical analysis of enzyme kinetic data. *Techniques in the life sciences, Biochemistry* vol. B1/11, Techniques in protein and enzyme biochemistry - part 2. pp B113/1-41, Amsterdam: Elsevier/North-Holland Biomedical Press, 1985.

Henley, J.P. & Sadana, A. "Categorization of enzyme deactivations using a series-type mechanism". *Enzyme and Microbial Technology*, 7, 50-60, 1985.

Koshland, D.E. Jr. "The comparison of non-enzymic and enzymic reaction velocities". *Journal of Theoretical Biology*, 2, 75-86, 1962.

Michaelis, L. & Menten, M.L. "The kinetics of invertin action". *Biochemische Zeitschrift*, 49, 333-69, 1913.

Bergmeyer, H.U. ed. *Methods of Enzymatic Analysis* 3rd edn, New York: Verlag Chemie, Academic Press, 1974.

Hall, E.A.H. "The developing biosensor arena". *Enzyme and Microbial Technology* 8, 651-657, 1986.

Kricka, L.J. & Thorpe, G.H.G. "Immobilised enzymes in analysis". *Trends in Biotechnology*4, 253-258, 1986.

Lowe, C.R. "Biosensors". *Trends in Biotechnology* 2, 59-64, 1984.

North, J.R. "Immunosensors: antibody-based biosensors". *Trends in Biotechnology* 3, 180-186, 1985.

Turner, A.P.F. "Biosensors: Principles and potential". In *Chemical aspects of food enzymes,* ed. A.T.Andrews, pp 259-270, London: Royal Society of Chemistry, 1987.

_____________, Karube, I. & Wilson, G.S. eds. *Biosensors: Fundamentals and applications,* Oxford, U.K: Oxford University Press, 1987.

Laane, C., Boeren, S., Vos, K. & Veeger, C. "Rules for optimization of biocatalysts in organic solvents". *Biotechnology and Bioengineering* , 30, 81-87, 1987.

Lilly, M.D. "Two-liquid-phase biocatalytic reactions". *Journal of Chemical Technology and Biotechnology,* 32, 162-169, 1982.

Luisi, P.L. & Laane, C. "Solubilization of enzymes in apolar solvents via reverse micelles". *Trends in Biotechnology,* 4, 153-160, 1986.

Nilsson, K.G.I. "Enzymatic synthesis of oligosaccharides". *Trends in Biotechnology,* 6, 256-264, 1988.

Zaks, A., Empie, M. & Gross, A. "Potentially commercial enzymatic processes for the fine and specialty chemical industries". *Trends in Biotechnology,* 6, 272-275, 1988.

_____________. & Klibanov, A.M. "Enzymic catalysis in organic media at 100°C". *Science,* 224, 1249-1251, 1984.

_____________. "Enzyme-catalyzed processess in organic solvents". *Proceedings of the National Academy of Sciences,* 82, 3192-3196, 1985.

Index